Ready For 2nd Grade

Math

A Summer Learning Activity Workbook For Students Entering Grade 2 [Ages 7-8]

Written and edited by Matthew Bowling

Printed in the United States of America

This paper meets the requirements of ANSI/NISO Z39.48-1992 (Permanence of Paper).

Paperback
ISBN 13: 978-1-5167-3122-0
ISBN 10: 1-5167-3122-0

TABLE OF CONTENTS

Introduction

You did it! You finished 1st grade! That is so exciting and awesome!

You are about to head into 2nd grade now, and that may be a little scary. This workbook is here to help you feel ready and confident as you tackle this new grade level. You've got this!

This workbook is organized into chapters. Each chapter will review important 1st grade skills and introduce new ideas you'll see in 2nd grade. Each chapter starts with a short lesson to refresh your memory or even introduce new things before you take on the activities for each day. You can always go back to that page for help when doing an activity page.

The workbook is made to be completed over 10 weeks, with each day taking no more than 15 minutes. That is just enough time to practice without feeling overwhelmed. Learning is an adventure! As you work through the chapters, the activities will slowly become trickier, but that just means your brain is growing stronger! This will help prepare you for bigger thinking and new tasks you will see in 2nd grade.

You might work with a grown-up sometimes, or you might complete much of it on your own. Either way, you're building skills, confidence, and a love for learning. Get ready, because 2nd grade is going to be a great year!

Chapter 1: Counting & Number Sense

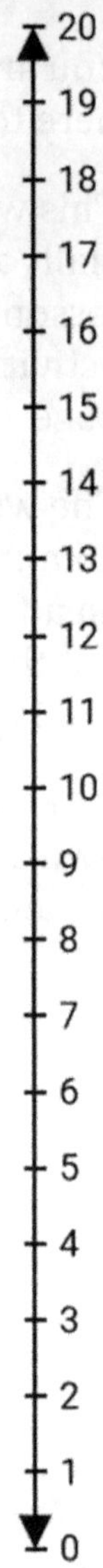

Counting is a process that determines the number or amount of objects in a set. **Numbers** are standard unique symbols that hold value. To count, one must rely on numbers and the understanding of the number sequence. The **number sequence** refers to the order in which numbers are arranged. **Number lines** are visuals that are used to display the correct number sequence on a line. These lines go on forever in both directions, so arrows are placed on either end of the line. Number lines can be displayed vertically or horizontally as shown below and on the side.

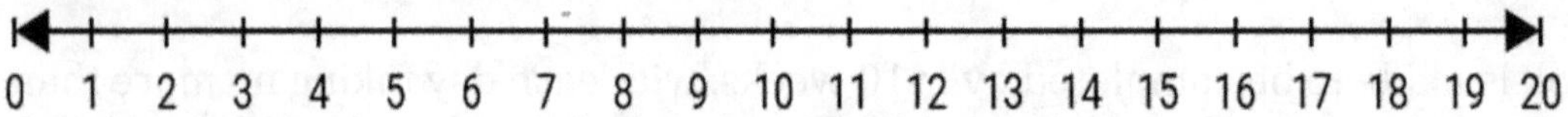

When counting it is important to remember that each object in a set must be counted exactly once, and the sequence must always stay the same. The last number spoken while counting objects represents how many total items are in the set. Also, the order or direction used to count the objects does not change the total number of items in the set. If the set hasn't changed, the number in the set does not have to be recounted each time. If there are no objects in a set, that means there are zero objects. Zero simply means none.

It is also helpful to count in larger increments. This is referred to as **skip counting**. Counting by twos, fives, tens, twenties, or hundreds are all examples of skip counting. An example of counting by fives would be 5, 10, 15, 20, 25, 30, and so on.

Numbers can also be written as words instead of symbols. For example, 1 is written as one and 55 is written as fifty-five.

Numbers can be broken apart and composed based on their **place value**. For example, the number 129,548 can be expressed in **expanded form** as 100,000 + 20,000 + 9,000 + 500 + 40 + 8. You should also be comfortable going from expanded form to standard form. For example, what number is represented by 30,000 + 400 + 80 + 2? Notice 3 groups of ten-thousands, 0 groups of thousands, 4 groups of hundreds, 8 groups of tens, and 2 groups of ones. This becomes 30,482.

Sometimes you will be asked to compose and decompose numbers that are presented in **picture** or **model** form. This usually includes base-10 blocks. For example, consider the model below. How many groups of thousands, hundreds, tens, and ones are there? There is 1 group of 1,000, 2 groups of 100, 4 groups of 10, and 7 ones. This forms the number 1,247.

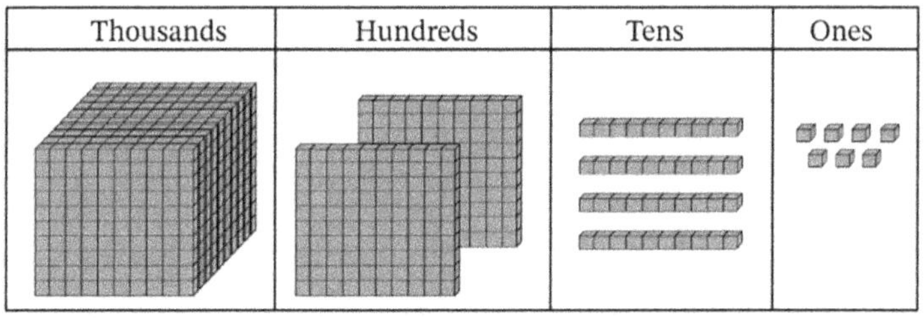

Two important symbols are the **inequality symbols** for **greater than** (>) and **less than** (<). For example, $45 > 35$ means 45 is greater than 35, and $26 < 62$ means 26 is less than 62. We can now build on this skill to compare and order whole numbers up to 100,000. The important thing to remember is place value. Remember, when comparing numbers, start by comparing the **farthest left digit**. For example, when comparing 456 and 453 start by comparing the hundreds place.

$\underline{4}56$ and $\underline{4}53$

In this case, both values are 4. Since they are the same, move to the tens place.

$4\underline{5}6$ and $4\underline{5}3$

In this case, both values are 5. Since they are the same, move to the ones place.

$45\underline{6}$ and $45\underline{3}$

6 is more than 3, so 456 is greater than 453.

This can be expressed using an inequality symbol: $456 > 453$.

Rounding is used to change a number to a nearby number that is easier to work with. To round, look at the digit after the place you want to round to. For example, if you are asked to round to the nearest tens place, look at the ones place. If the digit is 0, 1, 2, 3, or 4, the number in the place you are rounding to stays the same. This is referred to as rounding down. For example, 342 rounded to the nearest tens place is 340. If the digit is 5, 6, 7, 8, or 9, the number in the place you are rounding to goes up to the next value. This is referred to as rounding up. For example, 768 rounded to the nearest tens place is 770.

A **pattern** is something that repeats again and again in the same way or follows a specific rule. For example, the following pattern is skip counting by 2 and using the colors blue and then yellow: 2, blue, yellow, 4, blue, yellow, 6, blue, yellow.

Complete one worksheet each day in this chapter. Each day should take no longer than 15 minutes to complete.

DAY 1
PLACE VALUE

Draw an X over the third and twelfth spoons.

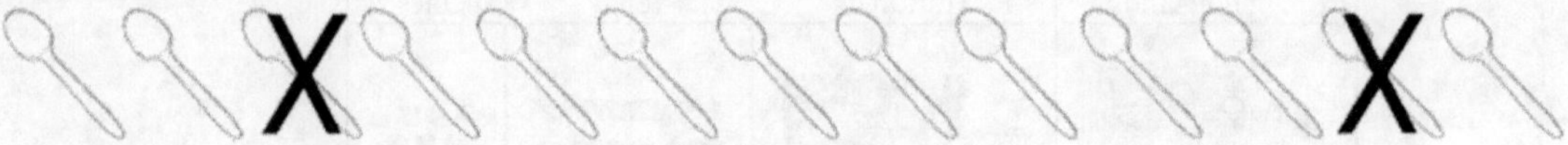

Draw an X over the fourth, ninth, and thirteenth boxes.

Draw an X over the eighth, tenth, and fourteenth lamps.

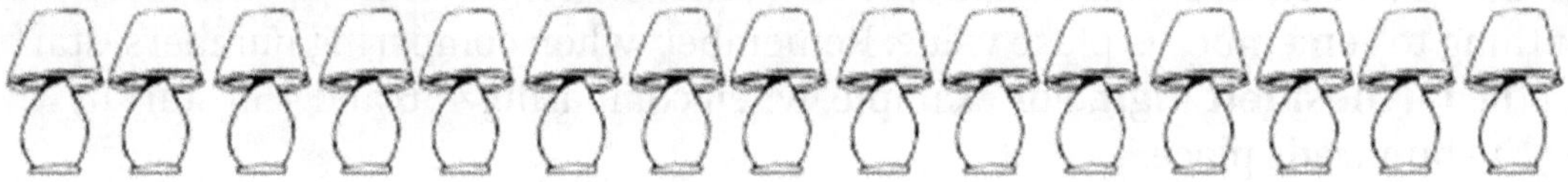

Draw an X over the first, eleventh, and fifteenth bags.

Draw an X over the second, fifth, and ninth objects.

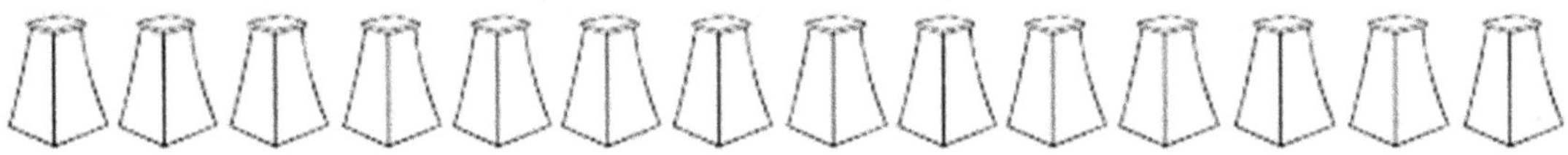

Draw an X over the fourth, seventh, and thirteenth globes.

Day 2

Greater Than, Less Than, or Equal: 2 Digit Numbers

Compare these 2 digit numbers and answer if each set is greater, less than, or equal.

1.	22	<	78	15.	22		57
2.	91		41	16.	22		45
3.	47		99	17.	22		22
4.	65		11	18.	78		78
5.	99		63	19.	42		24
6.	71		80	20.	31		33
7.	39		93	21	63		63
8.	10		10	22	88		25
9.	17		71	23	74		77
10.	49		77	24	22		12
11.	83		28	25	32		81
12.	22		37	26	98		99
13.	13		57	27	75		75
14.	93		52	28	22		11

DAY 3

ROUNDING TO THE TENS

Round each number to the nearest tens place.

1.	39	40	15.	14	______
2.	18	______	16.	95	______
3.	89	______	17.	77	______
4.	22	______	18.	13	______
5.	27	______	19.	58	______
6.	56	______	20.	21	______
7.	78	______	21	63	______
8.	11	______	22	34	______
9.	69	______	23	84	______
10.	72	______	24	90	______
11.	29	______	25	17	______
12.	81	______	26	84	______
13.	99	______	27	72	______
14.	76	______	28	19	______

ROUNDING TO THE NEAREST HUNDREDS

Round each number to the nearest hundreds place.

1.	399	400	15.	754	______
2.	452	______	16.	329	______
3.	124	______	17.	285	______
4.	695	______	18.	205	______
5.	230	______	19.	912	______
6.	425	______	20.	879	______
7.	475	______	21	325	______
8.	205	______	22	132	______
9.	244	______	23	897	______
10.	111	______	24	289	______
11.	587	______	25	475	______
12.	299	______	26	441	______
13.	199	______	27	654	______
14.	787	______	28	321	______

Day 4

Arrange Numbers from Smallest to Largest

Arrange the numbers in order from smallest to largest.

1.		2.		3.	
188	2	879	______	200	______
13	1	233	______	756	______
689	4	54	______	89	______
346	3	95	______	2	______

4.		5.		6.	
346	______	421	______	358	______
745	______	465	______	101	______
989	______	777	______	7	______
124	______	758	______	45	______

Arrange Numbers from Largest to Smallest

Arrange the numbers in order from largest to smallest.

1.		2.		3.	
123	4	564	______	54	______
676	2	2	______	123	______
879	1	99	______	124	______
333	3	612	______	53	______

4.		5.		6.	
202	______	99	______	125	______
457	______	45	______	141	______
589	______	54	______	223	______
2	______	470	______	789	______

Day 5

Understanding Place Value 1

The place value of a digit is shown by where it is in the number. Add the tens and ones to get your total.

1.	4	Tens	+	9	Ones	=	49
2.	2	Tens	+	0	Ones	=	
3.	6	Tens	+	5	Ones	=	
4.	5	Tens	+	5	Ones	=	
5.	9	Tens	+	1	Ones	=	

Understanding Place Value 2

Help Barry the bear solve these problems.

1.	600 + 90 + 5 = 695	6.	300 + 10 + 8 =
2.	100 + 70 + 3 =	7.	800 + 80 + 8 =
3.	400 + 60 + 1 =	8.	500 + 70 + 1 =
4.	200 + 20 + 9 =	9.	900 + 90 + 9 =
5.	900 + 30 + 4 =	10.	400 + 10 + 5 =

Understanding Place Value 3

Help Russell the alligator solve these problems.

1.	7	Hundreds	+	4	Tens	+	9	Ones	=	749
2.	4	Hundreds	+	2	Tens	+	0	Ones	=	
3.	5	Hundreds	+	1	Tens	+	5	Ones	=	
4.	1	Hundreds	+	5	Tens	+	5	Ones	=	
5.	3	Hundreds	+	9	Tens	+	1	Ones	=	

Day 6

Writing Numbers in Word Form

Write out each number.

1. 76 Seventy-Six
2. 34 __________
3. 59 __________
4. 13 __________
5. 99 __________
6. 85 __________
7. 24 __________
8. 42 __________
9. 435 __________
10. 211 __________
11. 147 __________
12. 654 __________
13. 174 __________
14. 697 __________
15. 387 __________
16. 311 __________

Day 7

Recognizing Patterns 1

Follow the directions in each section.

Write the numbers or letters that are next in order.

1. 1, ___, 3, A, 4, 5, 6, ___,
2. 4, 3, ___, 1, D, C, ___, A,
3. A, B, ___, D, E, ___, G, H,
4. 1, A, 2, B, ___, C, 4, ___

Write the number that is one less.

1.	___, 8	2.	___, 4	3.	___, 10	4.	___, 22
5.	___, 78	6.	___, 16	7.	___, 19	8.	___, 47
9.	___, 33	10.	___, 96	11.	___, 51	12.	___, 6
13.	___, 11	14.	___, 14				

Write the number that is one greater.

1.	___, 12	2.	___, 8	3.	___, 17	4.	___, 52
5.	___, 3	6.	___, 77	7.	___, 41	8.	___, 63
9.	___, 4	10.	___, 15	11.	___, 36	12.	___, 12
13.	___, 8	14.	___, 19				

DAY 8

RECOGNIZING PATTERNS 2: FILL IN THE NUMBER CHART

10	20		40			70		90	
110		130		150	160		180		
210			240		260			290	300
310	320			350			380		400
	420	430	440				480		
510		530		550	560		580		600
	620	630			660	670		690	
710			740	750					800
	820		840		860	870		890	
		930	940	950					1000

Chapter 2: Addition

Addition is when two or more sets of things are combined or joined together. This is also referred to as **composing**. The **sum** is the answer achieved after adding the sets. Similar to counting, unique symbols are used to indicate addition. The **plus sign** (+) represents addition, and the **equal sign** (=) is used to indicate the sum or answer. Numbers are used to represent the different sets of things that are being combined.

Addition can be done using numbers, visuals, and physical objects.

Place value plays an important role in addition and adding larger numbers. It is important to add values in each place value position with other values in the same place value position (i.e. ones with ones, tens with tens, etc.). When the sum of each place value is greater than 9, then an additional group is added to the next place value position. This is called **regrouping**. Look at the example below. When the numbers in the ones column add up to a number larger than 10, the first digit of that number carries over to the tens column.

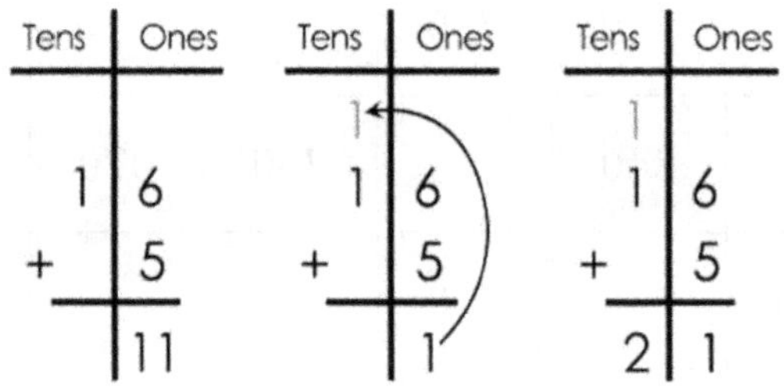

To do two digit addition, first add the numbers in the ones column, then add the numbers in the tens column.

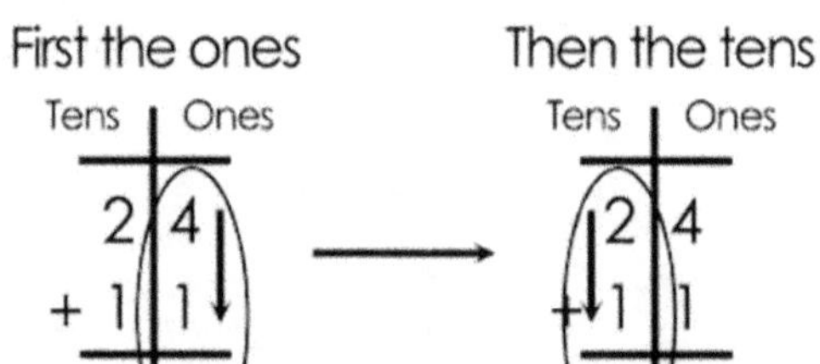

Commutative property means that even if the order of numbers is changed in an addition problem, the sum will always be the same.

Complete one worksheet each day in this chapter. Each day should take no longer than 15 minutes to complete.

Day 9

Single-Column Addition

Add the numbers, then circle the sum that is bigger in each set.

1.
$$\begin{array}{r} 7 \\ +\ 2 \\ \hline (9) \end{array} \qquad \begin{array}{r} 3 \\ +\ 5 \\ \hline 8 \end{array}$$

2.
$$\begin{array}{r} 5 \\ +\ 4 \\ \hline \end{array} \qquad \begin{array}{r} 2 \\ +\ 1 \\ \hline \end{array}$$

3.
$$\begin{array}{r} 5 \\ +\ 5 \\ \hline \end{array} \qquad \begin{array}{r} 3 \\ +\ 1 \\ \hline \end{array}$$

4.
$$\begin{array}{r} 1 \\ +\ 2 \\ \hline \end{array} \qquad \begin{array}{r} 8 \\ +\ 3 \\ \hline \end{array}$$

5.
$$\begin{array}{r} 5 \\ +\ 2 \\ \hline \end{array} \qquad \begin{array}{r} 9 \\ +\ 5 \\ \hline \end{array}$$

6.
$$\begin{array}{r} 6 \\ +\ 1 \\ \hline \end{array} \qquad \begin{array}{r} 7 \\ +\ 4 \\ \hline \end{array}$$

7.
$$\begin{array}{r} 2 \\ +\ 0 \\ \hline \end{array} \qquad \begin{array}{r} 4 \\ +\ 7 \\ \hline \end{array}$$

8.
$$\begin{array}{r} 3 \\ +\ 2 \\ \hline \end{array} \qquad \begin{array}{r} 8 \\ +\ 5 \\ \hline \end{array}$$

9.
$$\begin{array}{r} 1 \\ +\ 8 \\ \hline \end{array} \qquad \begin{array}{r} 9 \\ +\ 5 \\ \hline \end{array}$$

10.
$$\begin{array}{r} 2 \\ +\ 7 \\ \hline \end{array} \qquad \begin{array}{r} 6 \\ +\ 6 \\ \hline \end{array}$$

11.
$$\begin{array}{r} 8 \\ +\ 9 \\ \hline \end{array} \qquad \begin{array}{r} 4 \\ +\ 8 \\ \hline \end{array}$$

DAY 10

SINGLE-COLUMN ADDITION: GREATER THAN, LESS THAN OR EQUAL

Add the numbers, then decide if the first sum is greater than, less than, or equal to the second.

1.
$$\begin{array}{r} 2 \\ +\ 2 \\ \hline 4 \end{array} \quad \boxed{<} \quad \begin{array}{r} 3 \\ +\ 9 \\ \hline 12 \end{array}$$

2.
$$\begin{array}{r} 8 \\ +\ 1 \\ \hline \end{array} \quad \boxed{} \quad \begin{array}{r} 2 \\ +\ 7 \\ \hline \end{array}$$

3.
$$\begin{array}{r} 4 \\ +\ 8 \\ \hline \end{array} \quad \boxed{} \quad \begin{array}{r} 6 \\ +\ 5 \\ \hline \end{array}$$

4.
$$\begin{array}{r} 5 \\ +\ 2 \\ \hline \end{array} \quad \boxed{} \quad \begin{array}{r} 3 \\ +\ 6 \\ \hline \end{array}$$

5.
$$\begin{array}{r} 1 \\ +\ 9 \\ \hline \end{array} \quad \boxed{} \quad \begin{array}{r} 1 \\ +\ 6 \\ \hline \end{array}$$

6.
$$\begin{array}{r} 5 \\ +\ 8 \\ \hline \end{array} \quad \boxed{} \quad \begin{array}{r} 6 \\ +\ 5 \\ \hline \end{array}$$

7.
$$\begin{array}{r} 8 \\ +\ 1 \\ \hline \end{array} \quad \boxed{} \quad \begin{array}{r} 1 \\ +\ 1 \\ \hline \end{array}$$

8.
$$\begin{array}{r} 7 \\ +\ 2 \\ \hline \end{array} \quad \boxed{} \quad \begin{array}{r} 4 \\ +\ 5 \\ \hline \end{array}$$

9.
$$\begin{array}{r} 8 \\ +\ 8 \\ \hline \end{array} \quad \boxed{} \quad \begin{array}{r} 7 \\ +\ 2 \\ \hline \end{array}$$

10.
$$\begin{array}{r} 1 \\ +\ 2 \\ \hline \end{array} \quad \boxed{} \quad \begin{array}{r} 3 \\ +\ 3 \\ \hline \end{array}$$

11.
$$\begin{array}{r} 5 \\ +\ 7 \\ \hline \end{array} \quad \boxed{} \quad \begin{array}{r} 7 \\ +\ 7 \\ \hline \end{array}$$

Day 11

2-Digit Addition: Regrouping

Solve the problems below.

1. $\begin{array}{r} 16 \\ +\quad 5 \\ \hline 21 \end{array}$

2. $\begin{array}{r} 38 \\ +\quad 2 \\ \hline \end{array}$

3. $\begin{array}{r} 72 \\ +\quad 9 \\ \hline \end{array}$

4. $\begin{array}{r} 56 \\ +\quad 6 \\ \hline \end{array}$

5. $\begin{array}{r} 49 \\ +\quad 6 \\ \hline \end{array}$

6. $\begin{array}{r} 26 \\ +\quad 5 \\ \hline \end{array}$

7. $\begin{array}{r} 37 \\ +\quad 4 \\ \hline \end{array}$

8. $\begin{array}{r} 69 \\ +\quad 5 \\ \hline \end{array}$

9. $\begin{array}{r} 28 \\ +\quad 7 \\ \hline \end{array}$

10. $\begin{array}{r} 24 \\ +\quad 9 \\ \hline \end{array}$

11. $\begin{array}{r} 94 \\ +\quad 8 \\ \hline \end{array}$

12. $\begin{array}{r} 77 \\ +\quad 6 \\ \hline \end{array}$

Day 12

2-Digit Addition: Number Find

Solve the problems below, then circle the answer hidden in the scene.

41 + 8 = 49	64 + 4 = ___	92 + 7 = ___	45 + 5 = ___
24 + 3 = ___	37 + 2 = ___	16 + 1 = ___	26 + 9 = ___
57 + 7 = ___	19 + 4 = ___	21 + 7 = ___	85 + 9 = ___

Day 13

Commutative Property

Complete the problems by writing the missing numbers.

1.	$\underline{90} + 5 = 95$ $90 + \underline{5} = 95$		2.	__ + 6 = 77 71 + __ = 77
3.	__ + 70 = 80 10 + __ = 80		4.	__ + 7 = 15 8 + __ = 15
5.	__ + 25 = 46 21 + __ = 46		6.	__ + 74 = 98 24 + __ = 98
7.	__ + 13 = 20 7 + __ = 20		8.	__ + 13 = 64 51 + __ = 64
9.	__ + 87 = 92 5 + __ = 92		10.	__ + 20 = 29 9 + __ = 29
11.	__ + 3 = 65 62 + __ = 65		12.	__ + 3 = 34 31 + __ = 34

Day 14

2-Digit Addition

Solve the problems below.

1.	2.	3.	4.
24	73	24	64
+ 11	+ 23	+ 15	+ 20
35			

5.	6.	7.	8.
48	53	63	15
+ 51	+ 32	+ 21	+ 10

9.	10.	11.	12.
56	49	38	78
+ 32	+ 40	+ 61	+ 21

Day 15

3-Digit Single-Column Addition

Adding three or more numbers is just like adding two. Always remember to count them out.

	1.	2.	3.	4.	5.
4	5	1	9	2	6
5	4	1	4	2	1
+ 3	+ 7	+ 5	+ 6	+ 7	+ 8
12					

6.	7.	8.	9.	10.	11.
4	5	6	2	7	5
7	4	2	1	8	4
+ 1	+ 7	+ 9	+ 1	+ 8	+ 7

12.	13.
6	1
2	8
+ 3	+ 9

Day 16

4-Digit Single-Column Addition

Adding four or more numbers is a lot like adding two or three. Always remember to add them in sets.

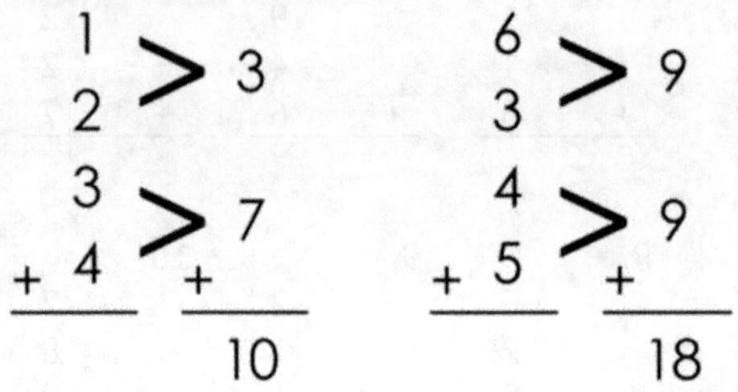

1.
$$\begin{array}{r} 4 \\ 5 \\ 3 \\ +\ 1 \\ \hline \end{array}$$

2.
$$\begin{array}{r} 8 \\ 1 \\ 2 \\ +\ 1 \\ \hline \end{array}$$

3.
$$\begin{array}{r} 5 \\ 3 \\ 3 \\ +\ 9 \\ \hline \end{array}$$

4.
$$\begin{array}{r} 1 \\ 2 \\ 2 \\ +\ 7 \\ \hline \end{array}$$

5.
$$\begin{array}{r} 7 \\ 4 \\ 2 \\ +\ 6 \\ \hline \end{array}$$

6.
$$\begin{array}{r} 5 \\ 6 \\ 6 \\ +\ 1 \\ \hline \end{array}$$

Chapter 3: Subtraction

Subtraction is when a set of things has a portion that is separated, removed, or taken away. This is also referred to as **decomposing**. The **difference** is the answer achieved after subtracting from the set. Similar to counting and addition, unique symbols are used to indicate subtraction. The **minus sign** (-) represents subtraction, and the **equal sign** (=) is used to indicate the difference or answer. Numbers are used to represent the different amounts.

Subtraction can be done using numbers, visuals, and physical objects.

Place value plays an important role in subtraction and subtracting larger numbers. It is important to subtract values in each place value position with other values in the same place value position (i.e. ones with ones, tens with tens, etc.). In subtraction, when the top digit has two numbers and the bottom digit has one, the digit in the tens column just drops down into the total.

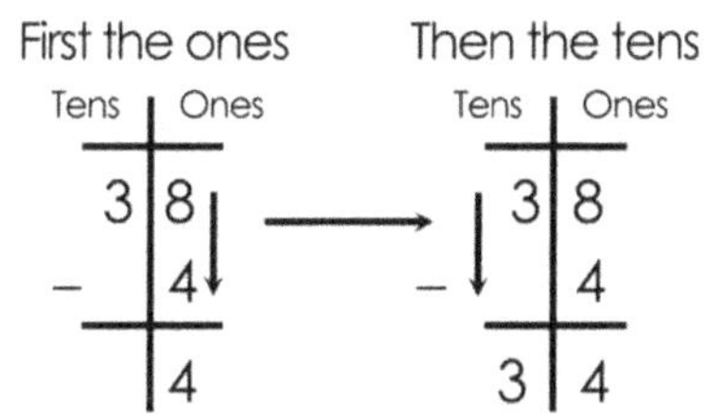

When doing two digit subtraction, subtract from the ones column first, then the tens column. If the top number in the ones column is smaller than the bottom, borrow a group of ten from the tens column.

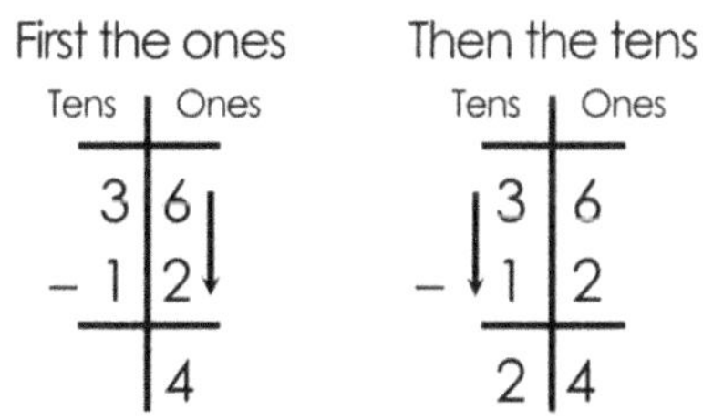

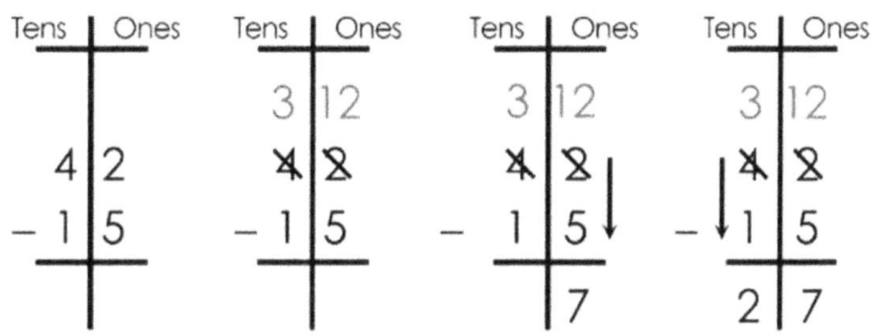

Complete one worksheet each day in this chapter. Each day should take no longer than 15 minutes to complete.

Day 17

Single-Column Subtraction: Greater Than, Less Than or Equal

Solve both equations in each set using subtraction. Then decide if the first difference in each set is greater than, less than, or equal to the second.

1. $\begin{array}{r}6\\-\ 2\\\hline 4\end{array}$ [<] $\begin{array}{r}9\\-\ 4\\\hline 5\end{array}$

2. $\begin{array}{r}7\\-\ 2\\\hline \end{array}$ [ ] $\begin{array}{r}3\\-\ 1\\\hline \end{array}$

3. $\begin{array}{r}2\\-\ 2\\\hline \end{array}$ [ ] $\begin{array}{r}8\\-\ 5\\\hline \end{array}$

4. $\begin{array}{r}3\\-\ 1\\\hline \end{array}$ [ ] $\begin{array}{r}5\\-\ 4\\\hline \end{array}$

5. $\begin{array}{r}8\\-\ 7\\\hline \end{array}$ [ ] $\begin{array}{r}2\\-\ 1\\\hline \end{array}$

6. $\begin{array}{r}10\\-\ 2\\\hline \end{array}$ [ ] $\begin{array}{r}5\\-\ 3\\\hline \end{array}$

7. $\begin{array}{r}3\\-\ 1\\\hline \end{array}$ [ ] $\begin{array}{r}9\\-\ 2\\\hline \end{array}$

8. $\begin{array}{r}6\\-\ 5\\\hline \end{array}$ [ ] $\begin{array}{r}8\\-\ 1\\\hline \end{array}$

9. $\begin{array}{r}2\\-\ 2\\\hline \end{array}$ [ ] $\begin{array}{r}3\\-\ 1\\\hline \end{array}$

10. $\begin{array}{r}6\\-\ 1\\\hline \end{array}$ [ ] $\begin{array}{r}8\\-\ 5\\\hline \end{array}$

11. $\begin{array}{r}10\\-\ 7\\\hline \end{array}$ [ ] $\begin{array}{r}5\\-\ 2\\\hline \end{array}$

Day 18

2-Digit Subtraction

Solve the subtraction problems below.

1. 75 - 2 = 73
2. 39 - 7
3. 56 - 5
4. 12 - 1
5. 23 - 3
6. 17 - 5
7. 29 - 5
8. 47 - 1
9. 86 - 6
10. 65 - 1
11. 88 - 5
12. 44 - 2

Day 19

2-Digit Subtraction: Number Find

Solve the problems below, then circle the answers hidden in the scene.

41 - 1 40	11 - 2	16 - 9	89 - 3
21 - 3	3 - 1	23 - 10	77 - 5
12 - 6	61 - 1	39 - 9	18 - 6

Day 20

2-Digit Subtraction

Solve the problems.

1.	2.	3.	4.
24	39	79	42
- 12	- 27	- 55	- 31
12			

5.	6.	7.	8.
68	17	59	97
- 43	- 15	- 35	- 11

9.	10.	11.	12.
26	39	28	64
- 16	- 21	- 25	- 42

Day 21

2-Digit Subtraction: Borrowing

Solve the problems.

1.	2.	3.	4.
61	54	84	91
- 24	- 19	- 46	- 73
37			

5.	6.	7.	8.
20	68	24	18
- 11	- 47	- 19	- 9

9.	10.	11.	12.
33	55	74	92
- 21	- 49	- 55	- 39

Day 22

Writing out 2-Digit Subtraction Problems

Write out these subtraction problems in columns and work them out.

1. 28-12

```
   28
-  12
-----
   16
```

2. 65-37

```
   65
-  37
-----
```

3. 69-51

```
   69
-  51
-----
```

4. 12-10

```
   12
-  10
-----
```

5. 85-72

```
   85
-  72
-----
```

6. 51-43

```
   51
-  43
-----
```

7. 30-27

```
   30
-  27
-----
```

8. 96-53

```
   96
-  53
-----
```

Day 23

Subtraction Word Problems

Use subtraction to solve.

1. George has three dogs. Sally has one dog. How many more dogs does George have than Sally? ___________

2. Emma had nine pieces of candy. She ate three. How many pieces of candy does she have left? ___________

3. David needs to run six miles to finish the race. He has run two so far. How many miles does he have left to run? ___________

4. The school day is seven hours. Mark has been at school for four hours. How many hours does he have left? ___________

5. Jimmy's dog weighed 80 pounds. He lost 17 pounds. How much does he weigh now? ___________

Chapter 4: Fractions & Graphing

Fractions are numbers that make up part of a whole. For example, the pizza, watermelon, and lemon are each sliced into equal pieces. The pizza has 6 equal pieces, the watermelon has 4 equal pieces, and the lemon has 8 equal pieces. These values represent the bottom number in a fraction, which represents the **total number of pieces**.

$$\frac{?}{6} \qquad \frac{?}{4} \qquad \frac{?}{8}$$

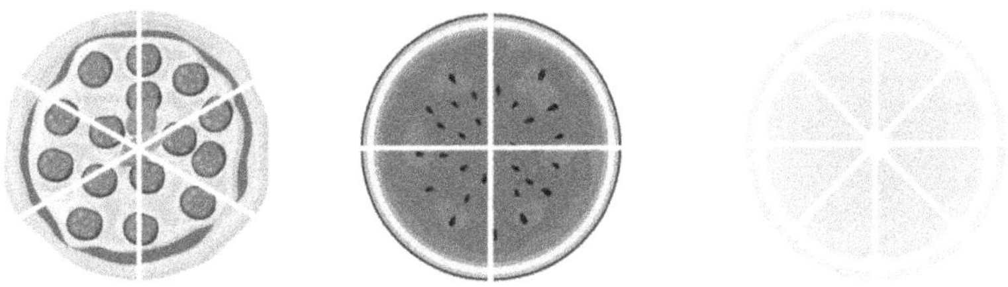

If 3 slices of each food are eaten, 3 would be the top number in each fraction. Some other examples are shown below.

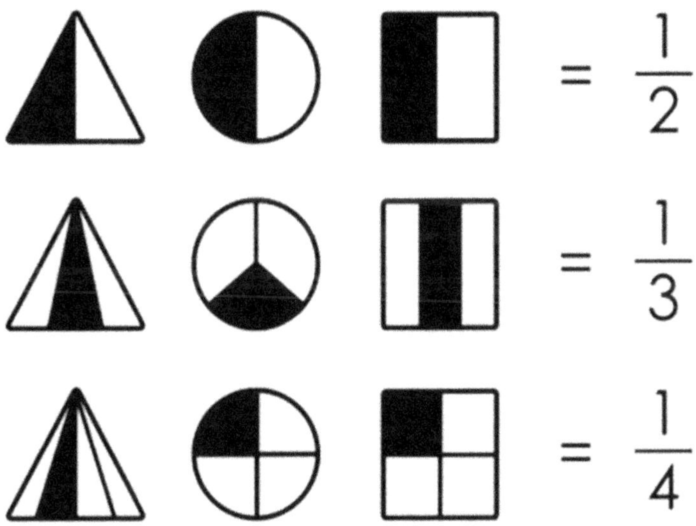

The shaded parts represent the top of the fractions. Again, the bottom numbers represent the total number of equal pieces each shape is broken up into.

Data refers to a collection of information. Data can be summarized using different graphical displays such as **bar graphs** and **pictographs**.

Complete one worksheet each day in this chapter. Each day should take no longer than 15 minutes to complete.

Day 24

Halving

Shade in half of each shape below.

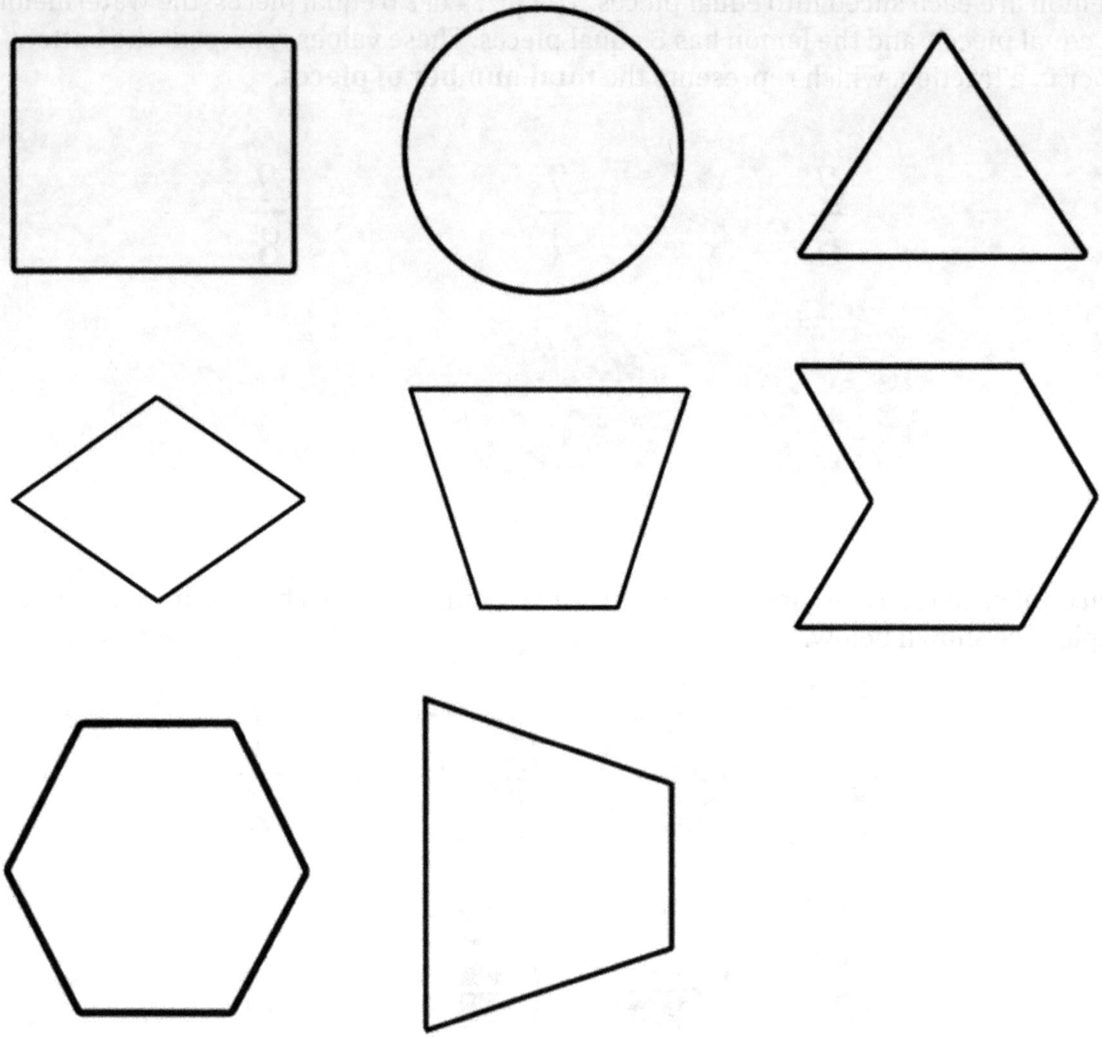

Day 25

Identifying Fractions

Color the shapes to match the fractions.

$\frac{1}{3}$ = 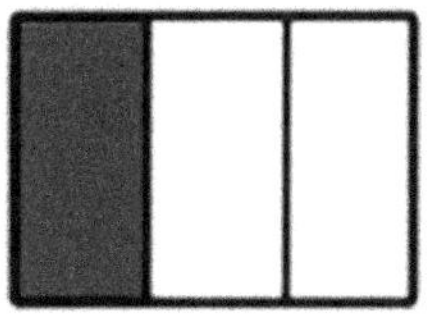

$\frac{2}{6}$ = 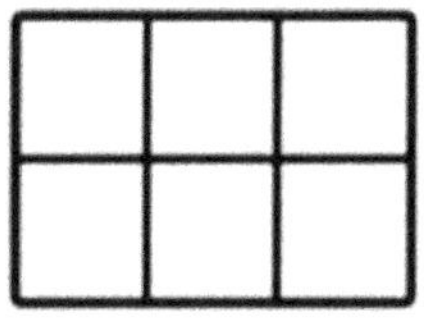

$\frac{1}{4}$ = 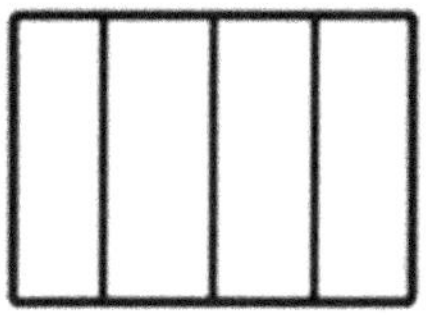

$\frac{5}{6}$ = 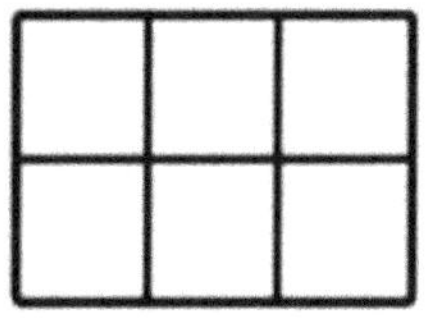

$\frac{2}{3}$ = 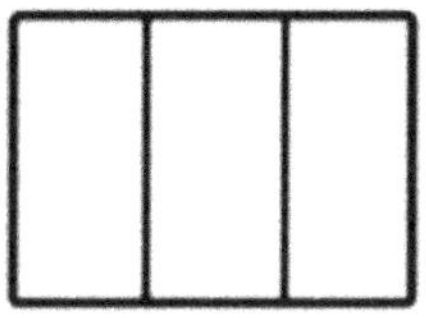

$\frac{7}{9}$ = 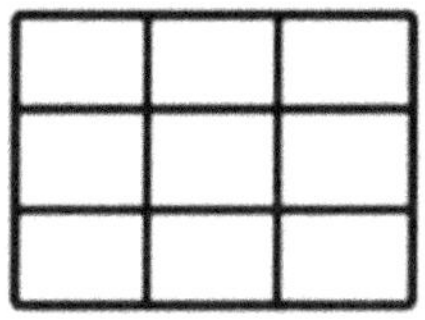

$\frac{3}{8}$ = 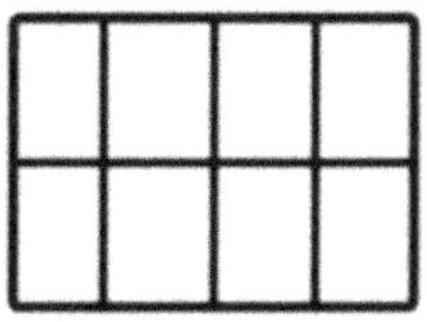

$\frac{1}{2}$ = 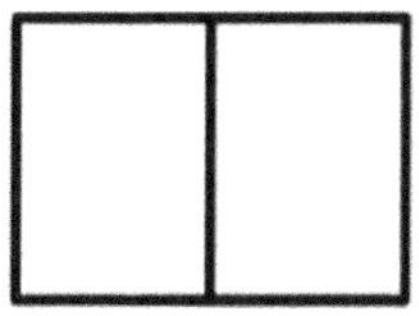

Day 26

Writing Fractions

Write the fractions in number form. The shaded part goes on top of the fraction.

	= $\frac{1}{2}$		=
	=	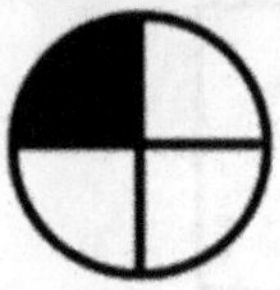	=
	=		=
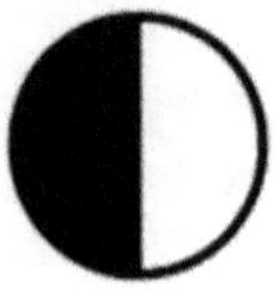	=		=

Day 27

Fraction Word Problems

Solve these fraction problems by coloring in the correct area.

1. Jimmy ate half of his sandwich. Color half of his sandwich.

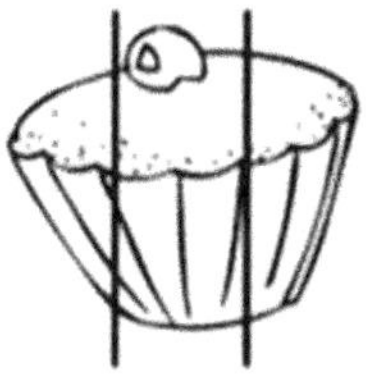

2. Amy can have one third of this cupcake. Color one third of her cupcake.

3. Penny ate one quarter of the pizza. Color one quarter of her pizza.

DAY 28

GRAPHING

Count the hamburgers in each row. Color the boxes to match the numbers.

Day 29

Graphing: Identifying Coordinates

To find places on a graph, start at 0, count across, then up. These are called coordinates.

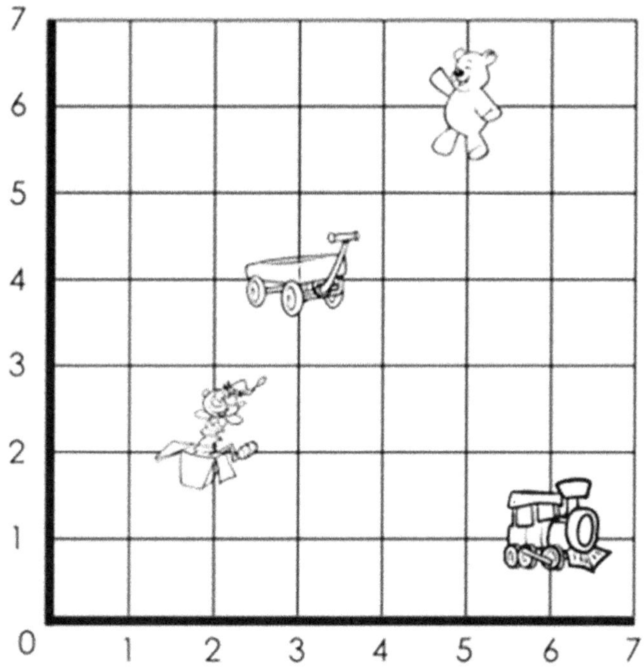

Write the coordinates for each toy in the blanks.

Across , Up	Across , Up	Across , Up	Across , Up
3 , 4	____ , ____	____ , ____	____ , ____

DAY 30

GRAPHING WORD PROBLEMS

Use the graph to answer the questions.

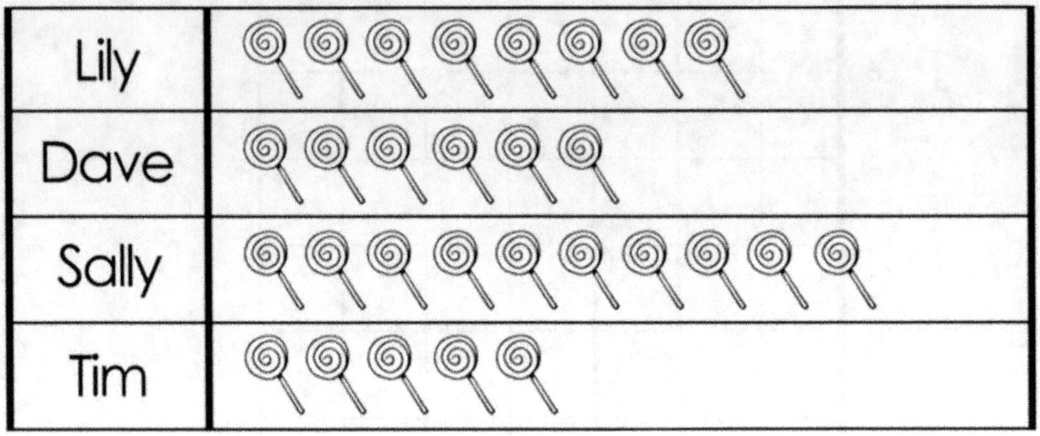

1. Lily has how many more pieces than Tim? ________
2. ________ has the most candy.
3. ________ has the least amount of candy.
4. Sally has how many more pieces of candy than Tim?
5. Write each kid's name and how many pieces of candy they have.

____

Chapter 5: Counting Money

Each coin and bill has its own value.

= 1 penny = 1¢ = $0.01

= 1 nickel = 5¢ = $0.05

= 1 dime = 10¢ = $0.10

= 1 quarter = 25¢ = $0.25

= 1 dollar = $1.00

= 5 dollars = $5.00

 = 10 dollars = $10.00

 = 20 dollars = $20.00

The value of a collection of coins can be found by adding the values according to place value. 100¢ is equal to one dollar ($1.00). Therefore, if there is a total of 130¢ in a collection of coins, there is actually a total of one dollar and 30 cents ($1.30).

Complete one worksheet each day in this chapter. Each day should take no longer than 15 minutes to complete.

Day 31

Counting Change

Write how many of the missing coin you need to buy each object.

	1	1		3
	2	1	1	
	2		1	2

Day 32

Cents and Dollars

Write in each box how many pieces of change you need to equal each bill.

How many  =  4

How many  = 

How many  = 

How many  = 

Day 33

Dollars and Larger Bills

Write in each box how many pieces of change you need to equal each bill.

How many 

= 

5

How many 

= 

How many 

= 

How many 

= 

Day 34

Dollars: Greater Than, Less Than or Equal

Compare the money groups. Write in the box which group is greater than, less than, or equal.

<

DAY 35

DOLLARS: MATCHING

Match the sets of change with the correct dollar amount.

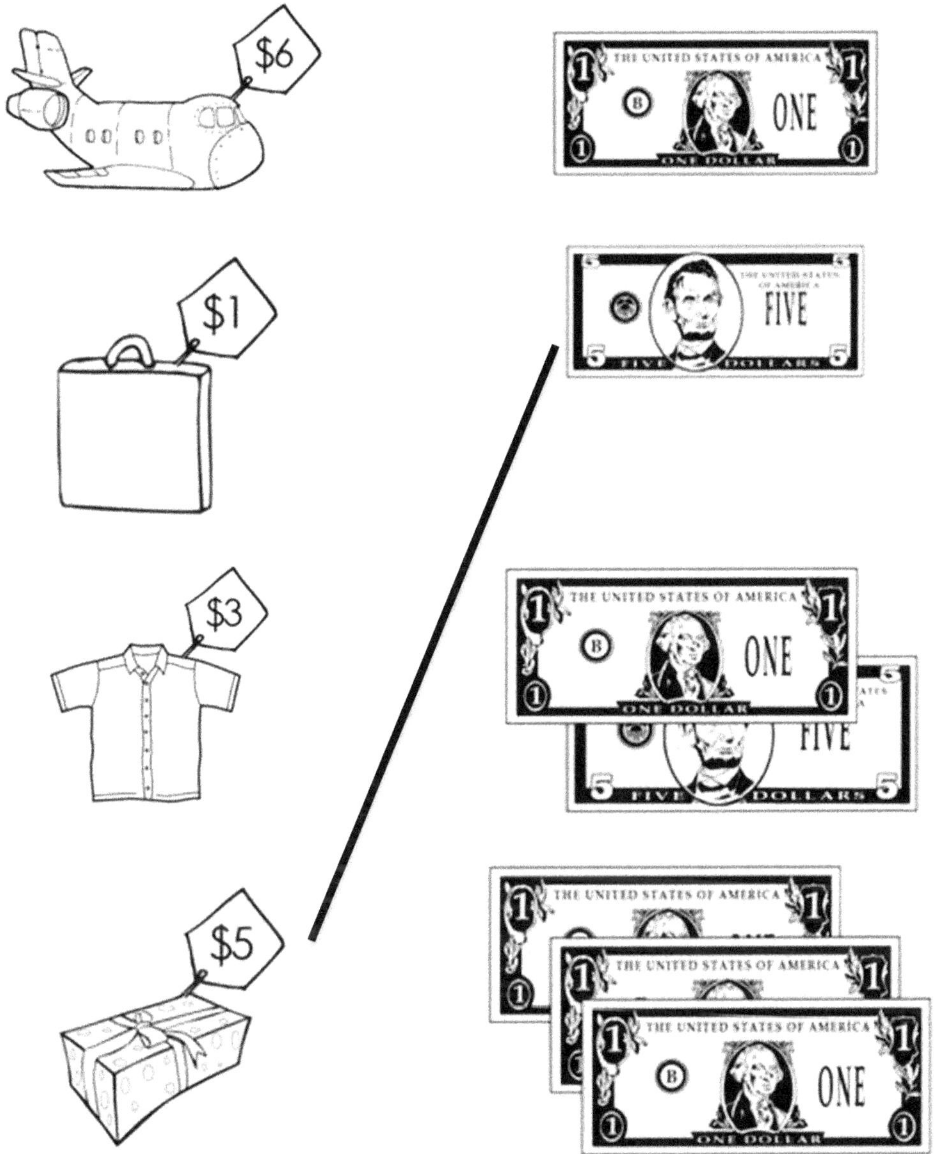

Chapter 6: Time

The clock below displays 4:00. The short hand, referred to as the **hour hand**, points to 4, and the long hand, referred to as the **minute hand**, points to 12 which represents 0 minutes.

Moving to the right around the clock starting at 12 is referred to as **clockwise**. In order to tell how many minutes have passed in each hour, the minute hand moves clockwise around the clock. Each number starting at 1 represents an interval of 5. For example, 1 represents 5 minutes, 2 represents 10 minutes, 3 represents 15 minutes, and so on. When the minute hand reaches 12, the number of minutes resets to zero, and the hour hand moves clockwise to the next number to represent the hour. It is also important to note that the hour hand moves slightly as minutes pass, and it will be closest to whatever hour mark is closest. For example, if the time is 4:50, the hour hand will be closer to 5 than 4. Likewise, if the time is 4:10, the hour hand will be closer to 4 than 5.

There are a total of 60 minutes in one hour. One-half of one hour is 30 minutes. One-quarter or one-fourth of one hour is 15 minutes. 45 minutes is three-quarters of an hour or one-quarter of an hour until the next hour since there are only 15 minutes left of the current hour.

Complete one worksheet each day in this chapter. Each day should take no longer than 15 minutes to complete.

Day 36

Identifying Time

What time is it? Write the answer under each clock.

3:00

3:30

DAY 37

WRITING TIME

Write the correct time in the clock on the right.

DAY 38

DRAWING TIME

Draw the correct time in the clock on the right.

DAY 39

TIME MATCHING

Draw a line from each clock on the left to match the clock on the right.

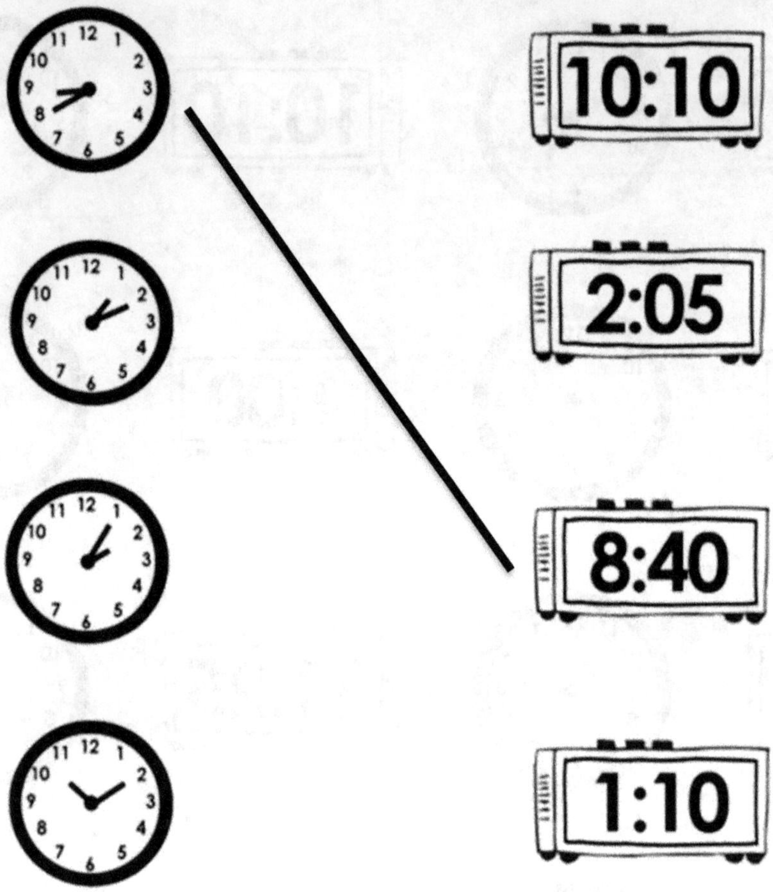

DAY 40

WRITING TIME PASSAGE

Write the correct time in the clocks on the right.

What time will it be in 3 hours and 30 minutes?

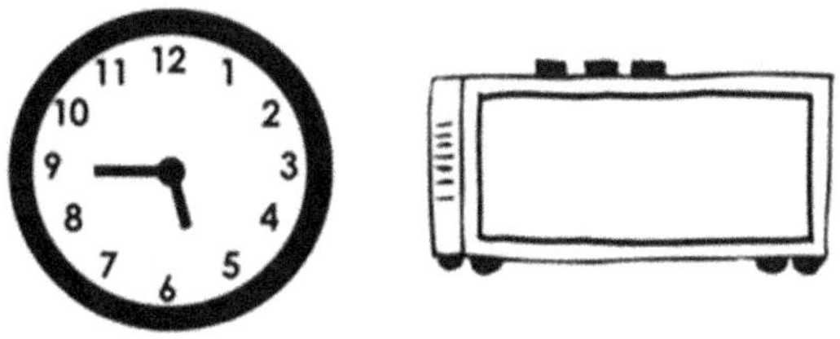

What time will it be in 6 hours and 30 minutes?

What time will it be in 1 hour and 45 minutes?

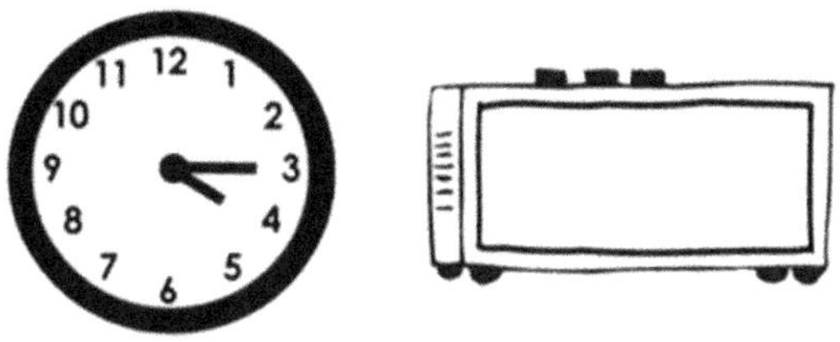

What time will it be in 3 hours and 30 minutes?

What time will it be in 3 hours and 45 minutes?

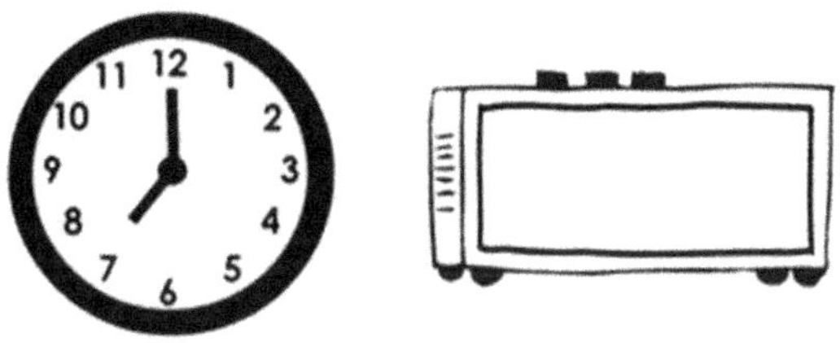

What time will it be in 2 hours and 30 minutes?

Chapter 7: Measurements

Measurements are used to determine how big, tall, long, heavy, or hot something is by comparing it to other things. Sometimes, objects are used to measure. Other times, precise measuring tools like a ruler or a scale are used. In America, the customary system of measurement is primarily used. Some of the basic units of measurement in this system include inches, feet, yards, miles, ounces, pounds, cups, and gallons. However, most other countries use a different system of measurement, and this type of measurement is often used in math and science in America as well. Some basic units of measurement in the metric system include meters, centimeters, kilometers, liters, and grams.

These are a few measuring tools that can be used.

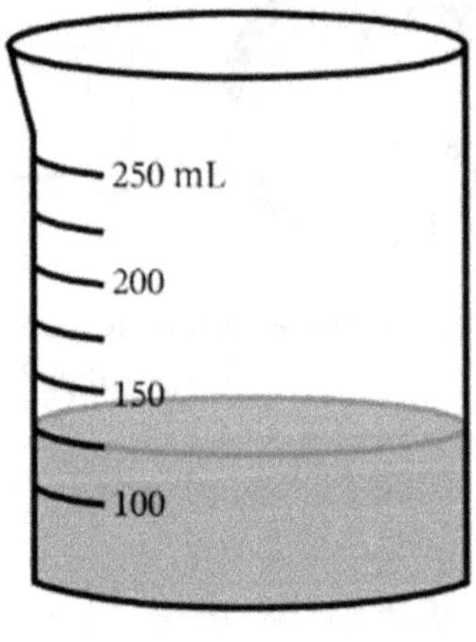

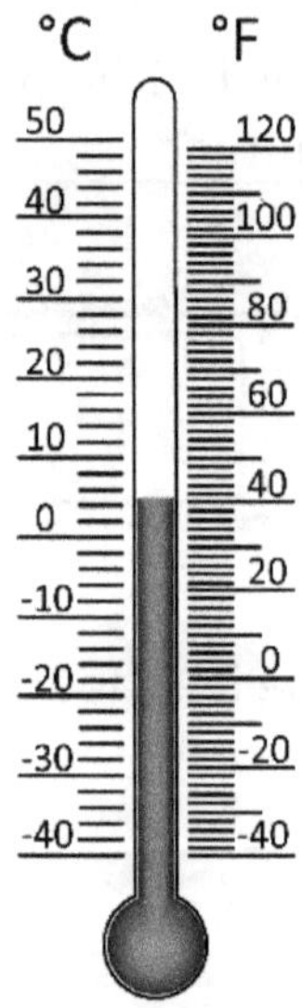

Complete one worksheet each day in this chapter. Each day should take no longer than 15 minutes to complete.

DAY 41

IDENTIFYING LIQUID MEASUREMENTS

Number the objects in order from the least to the most amount of liquid.

2 1 3 4

Day 42

Comparing Liquid Measurements

4 cups = 1 quart

Circle the set that holds more liquid.

 or 

 or 

 or 

 or 

Day 43

Identifying Warmer Temperatures

Circle the warmer temperature.

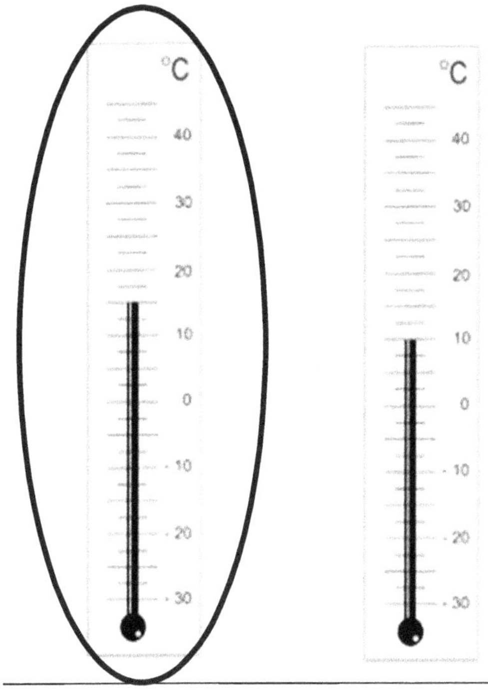

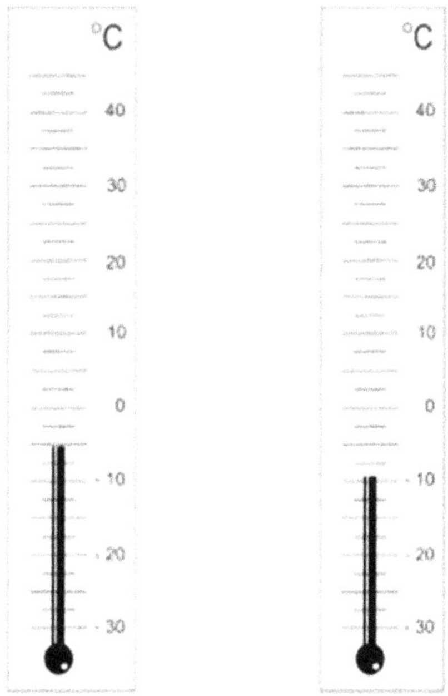

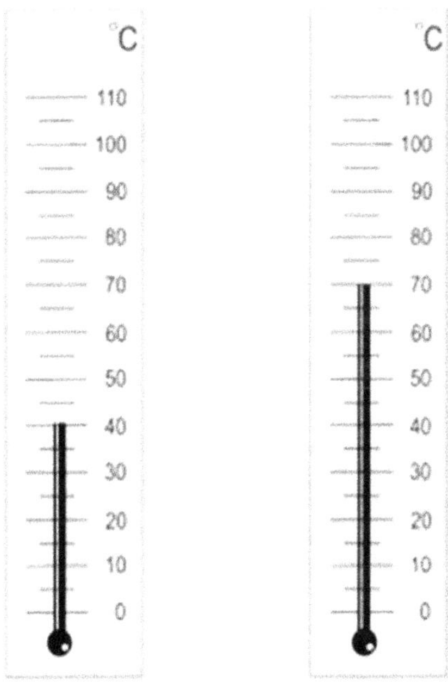

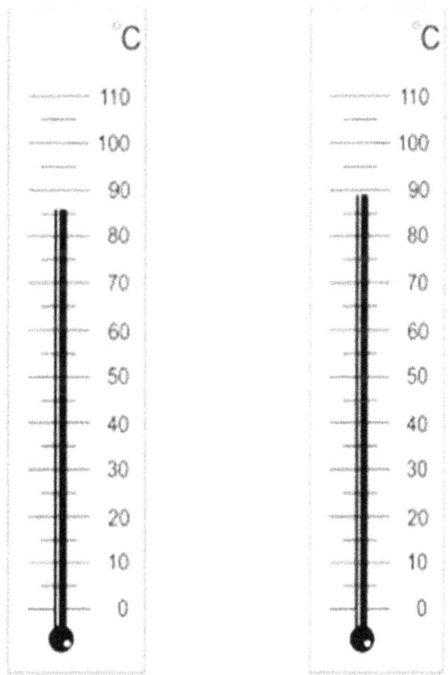

Day 44

Writing Temperature

Write the correct temperature in the blank.

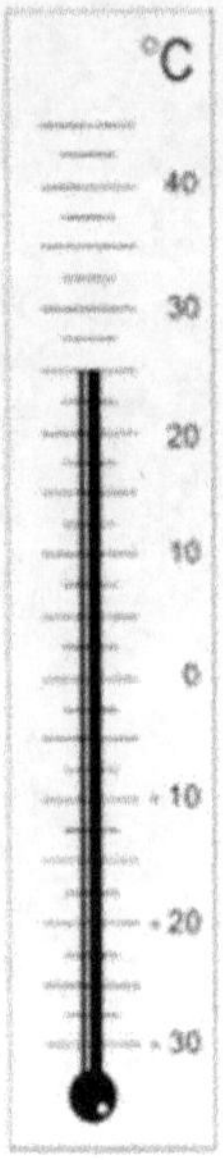

25°

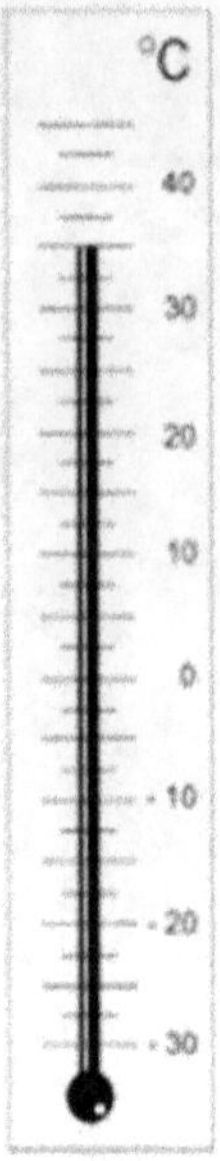

____

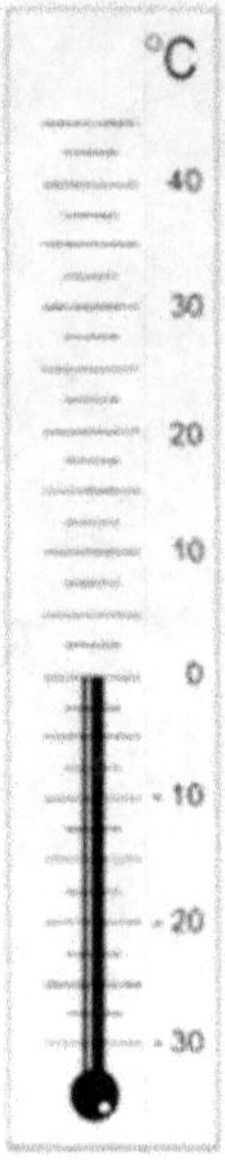

____

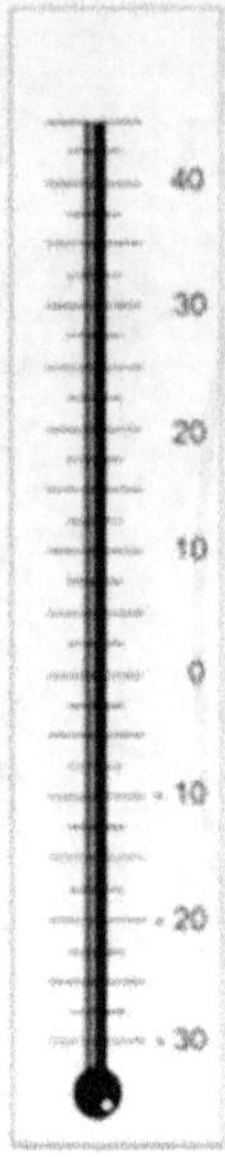

____

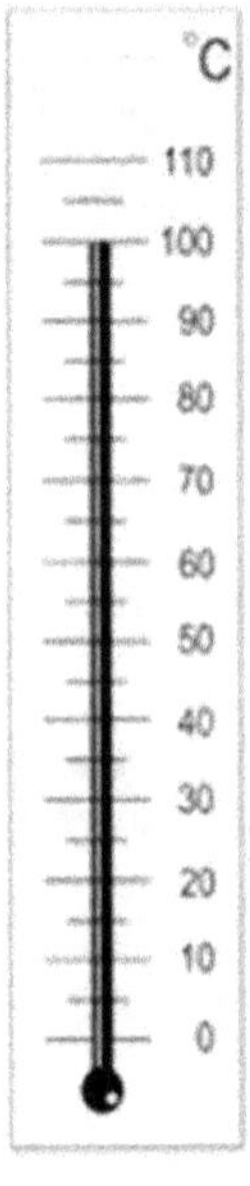

____

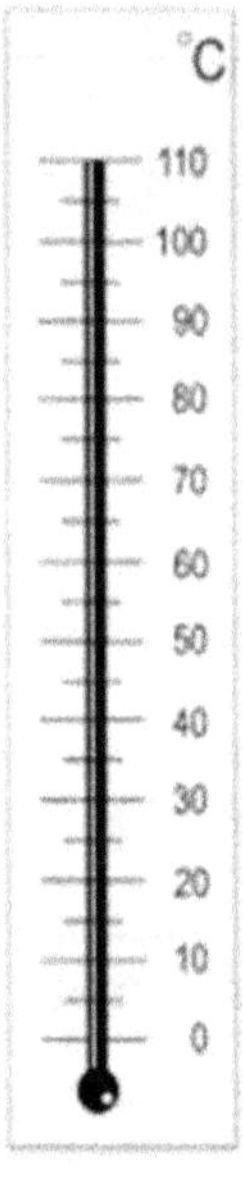

____

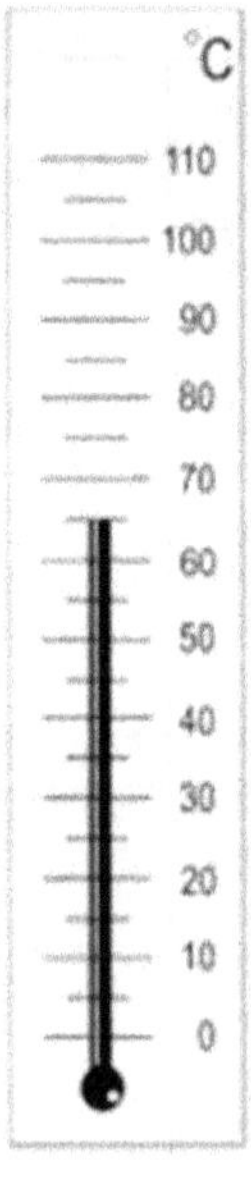

____

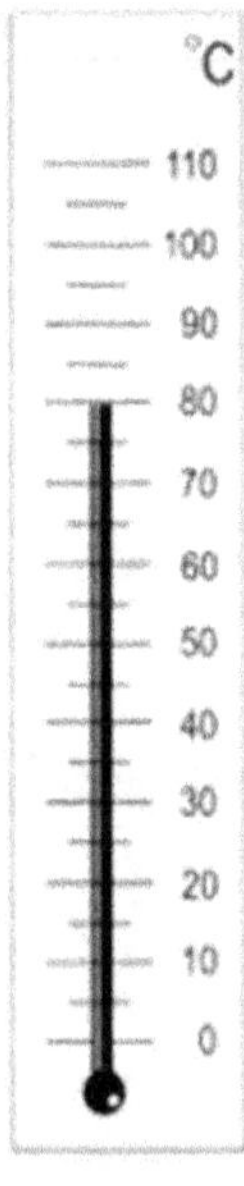

____

Day 45

Identifying Length in Centimeters

Write the height of each object in the box.

15 centimeters	centimeters	centimeters

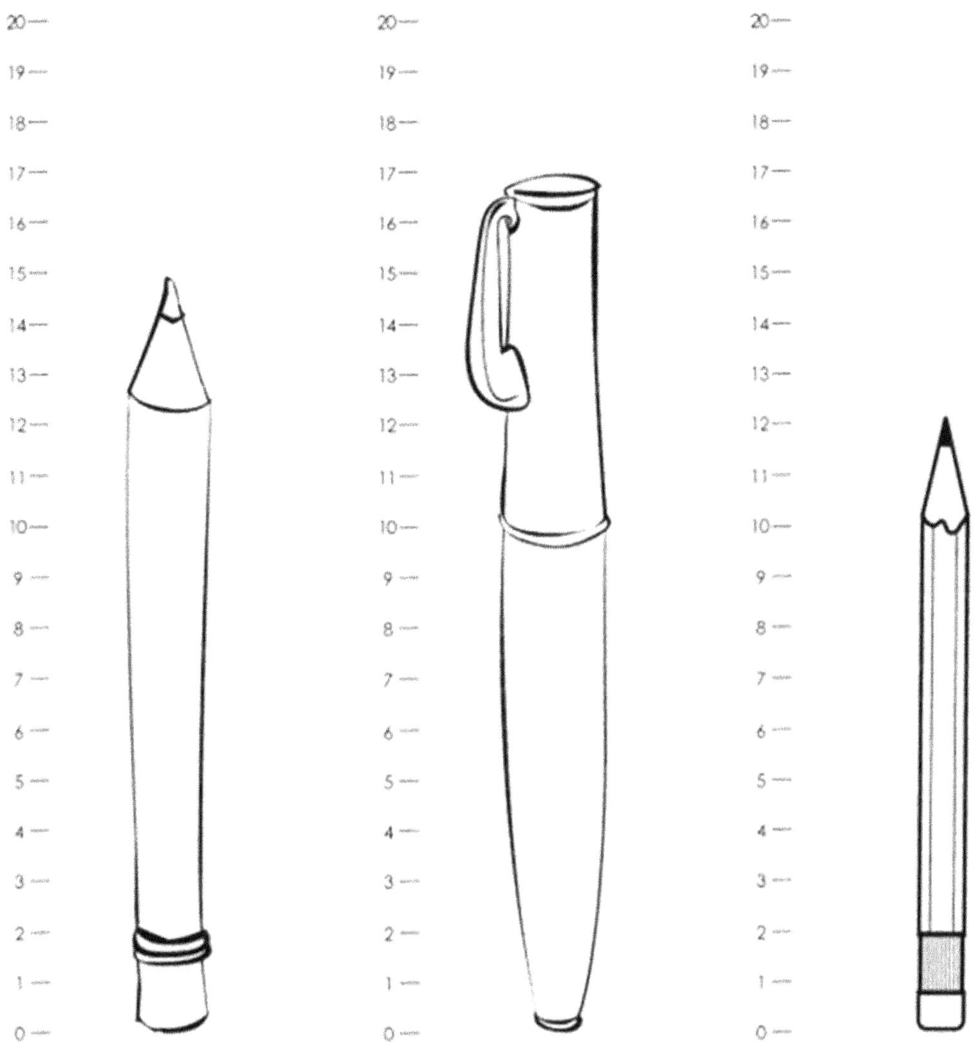

Chapter 8: Geometry

Two-dimensional shapes can be classified by the number of sides they have. Polygons are considered **regular polygons** if all the sides are the same length. For example, a regular pentagon has 5 sides, and all of the sides are the same length. Side lengths that are the same are considered **congruent**. For example, a square has 4 congruent side lengths.

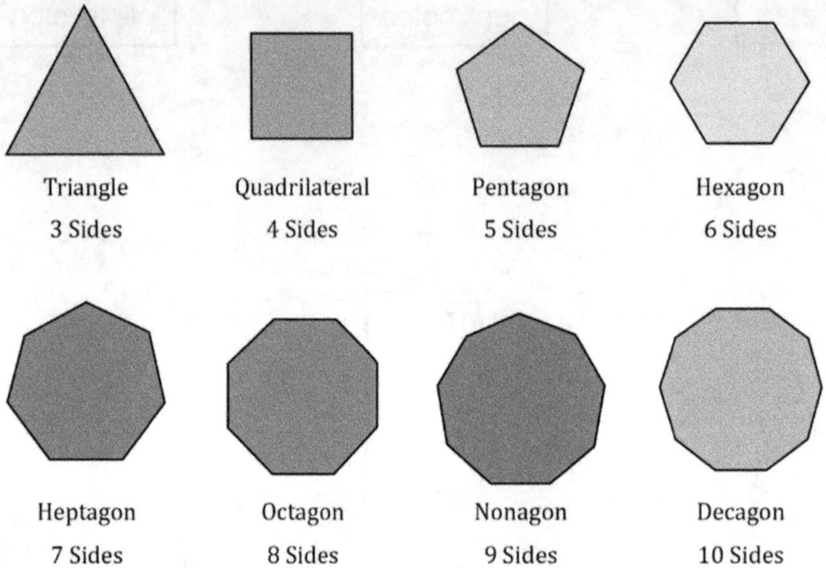

The most common three-dimensional figures include cones, cylinders, spheres, triangular and rectangular prisms, and cubes. When identifying and classifying three-dimensional figures, it is important to be aware of the number of faces, edges, and vertices. **Faces** are the flat surfaces, **edges** are where two faces come together, and **vertices** are like corners, where more than two faces come together. For example, a rectangular prism has 6 faces, 12 edges, and 8 vertices.

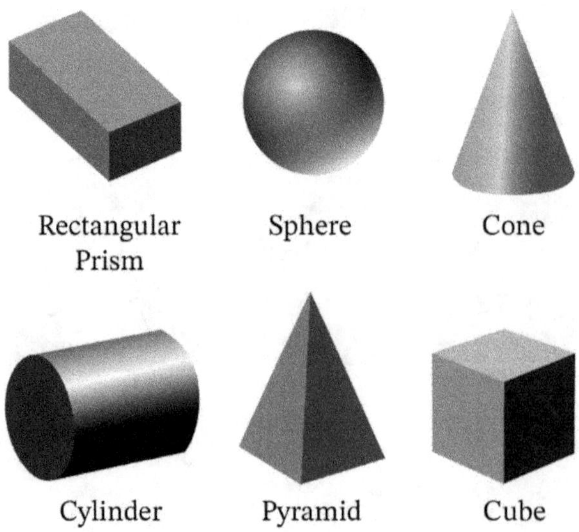

A line of symmetry is a line that divides a shape or an object into two equal and symmetrical parts. These parts can be folded together, and they will form a perfect match.

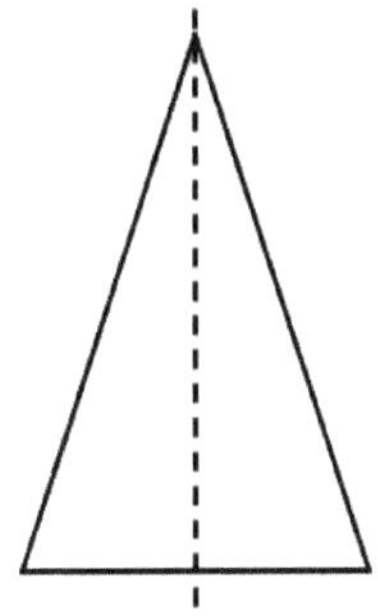

Complete one worksheet each day in this chapter. Each day should take no longer than 15 minutes to complete.

Day 46

Identifying Flat Shapes

Write the name of each shape.

 ______________

 ______________

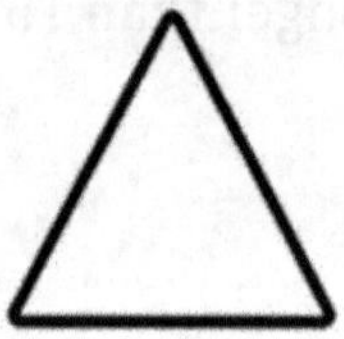 ______________

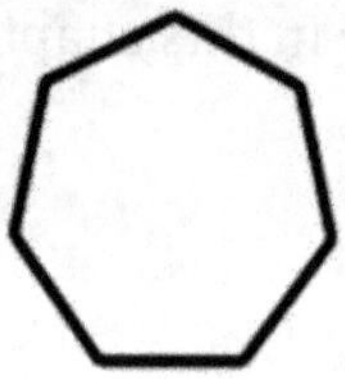 ______________

 ______________

 ______________

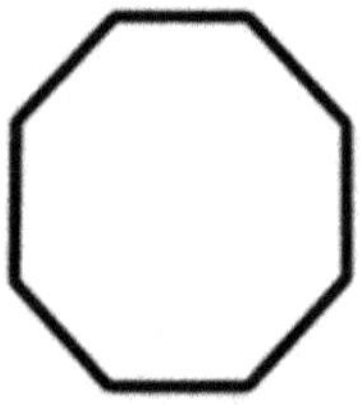 ______________

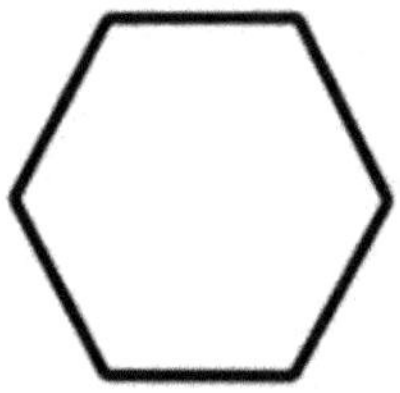 ______________

DAY 47

IDENTIFYING SOLID SHAPES

Write the name of each shape.

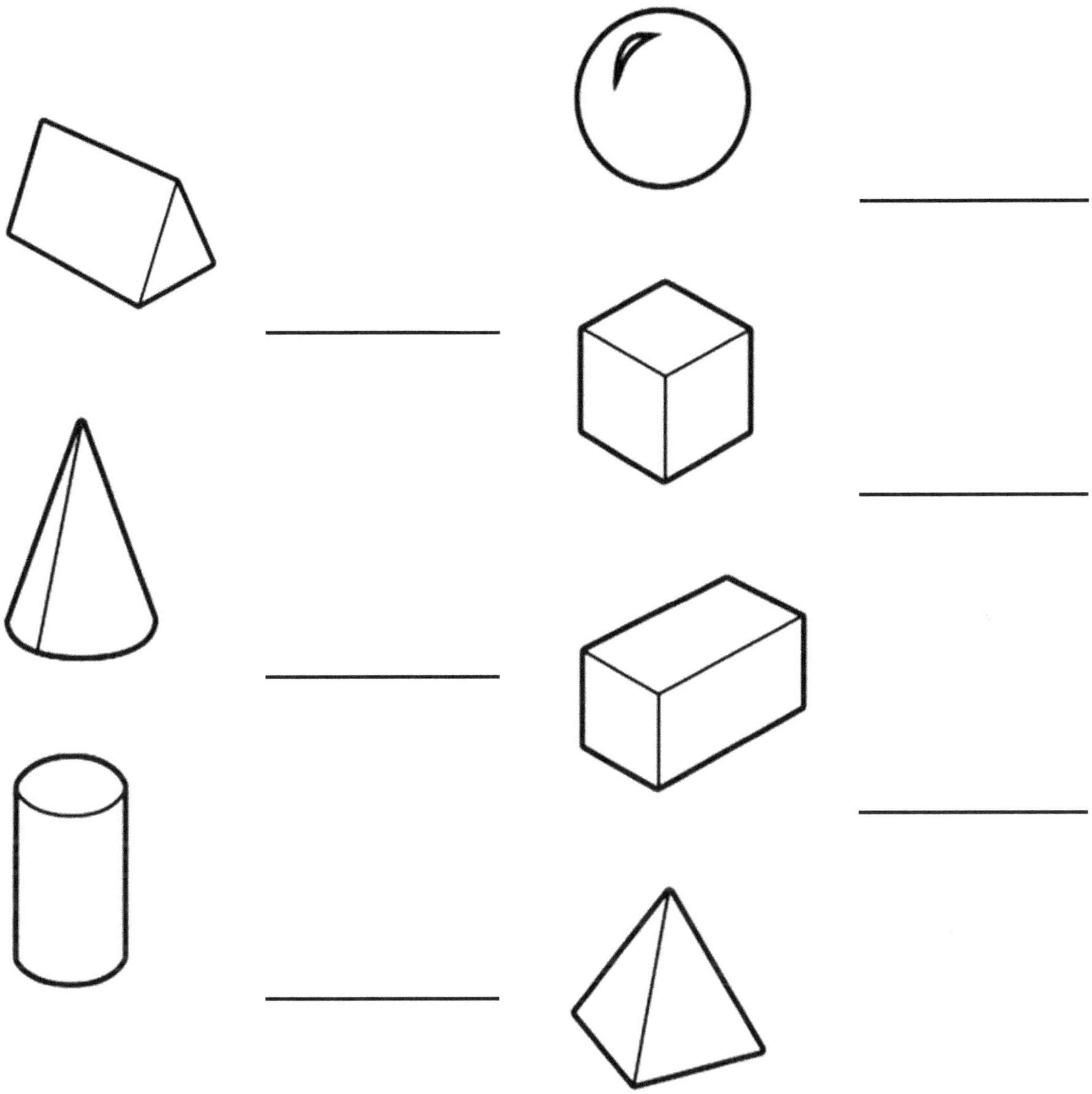

Day 48

Making New Shapes

Draw lines through each shape below to make new shapes.

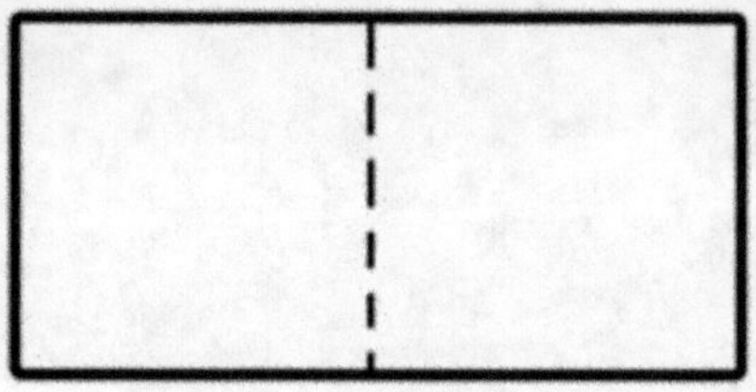

1. Make two squares.

2. Make a square and a triangle.

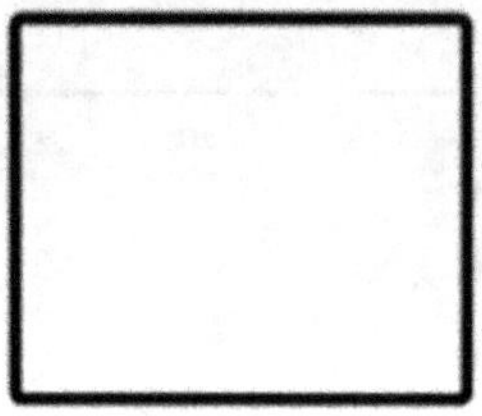

3. Make two triangles.

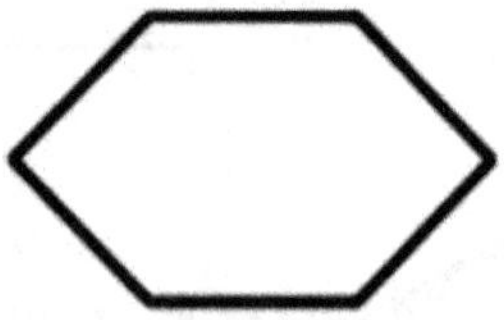

4. Draw two lines to make a square and two triangles.

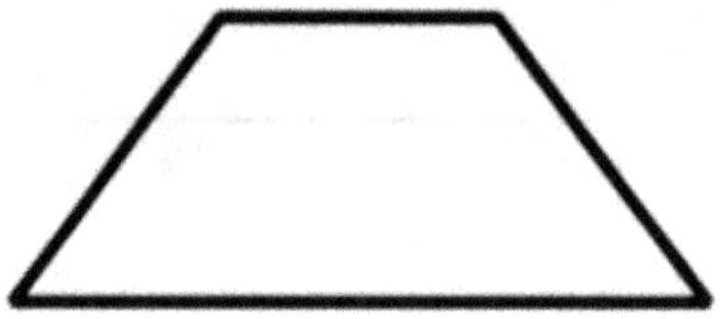

5. Draw two lines to make a square and two triangles.

DAY 49

COMPARING SOLID SHAPES

Circle the objects that are the same as the solid shapes in each row.

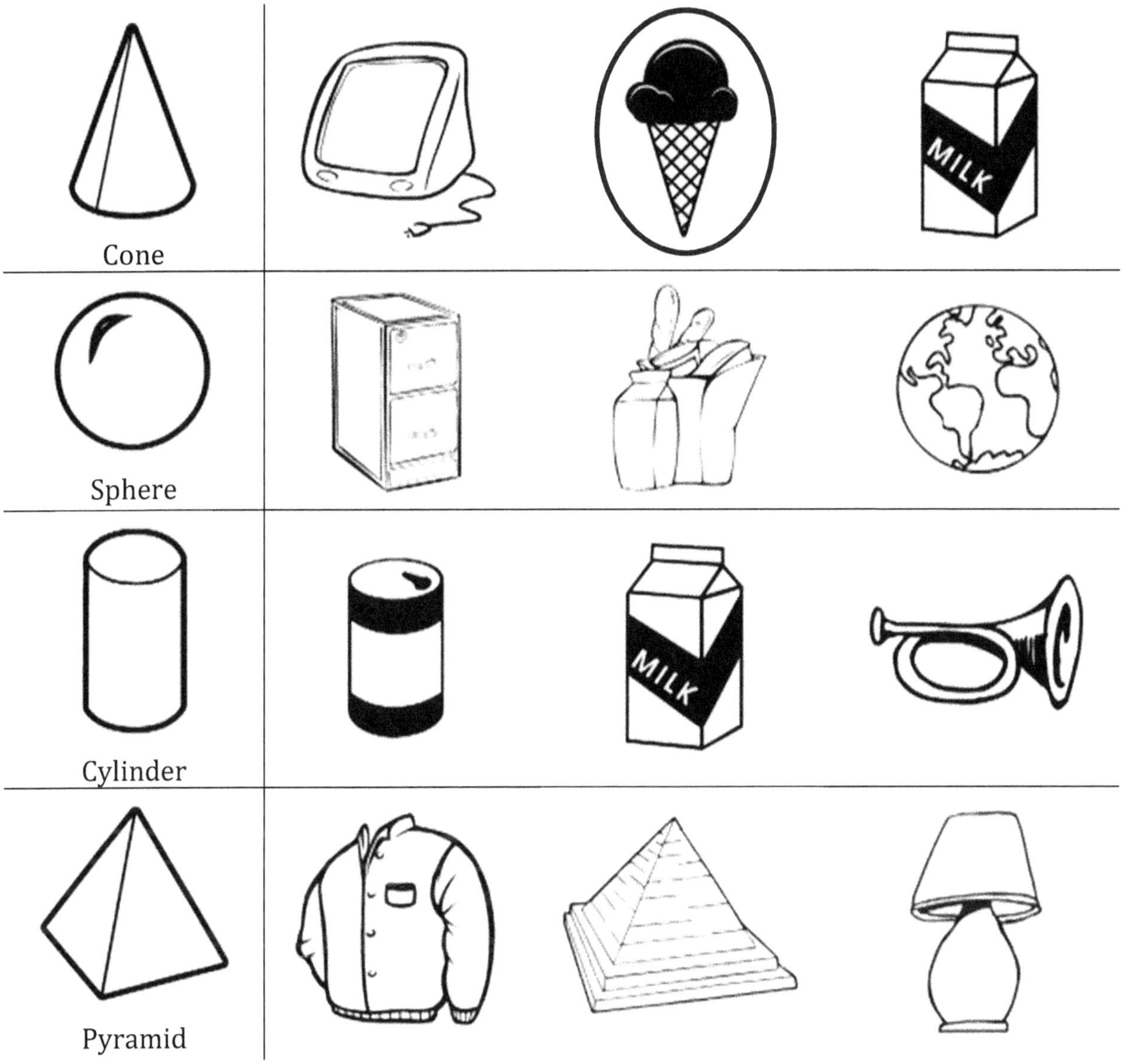

DAY 50
SYMMETRY

Finish drawing the other side of each shape to make both sides the same.

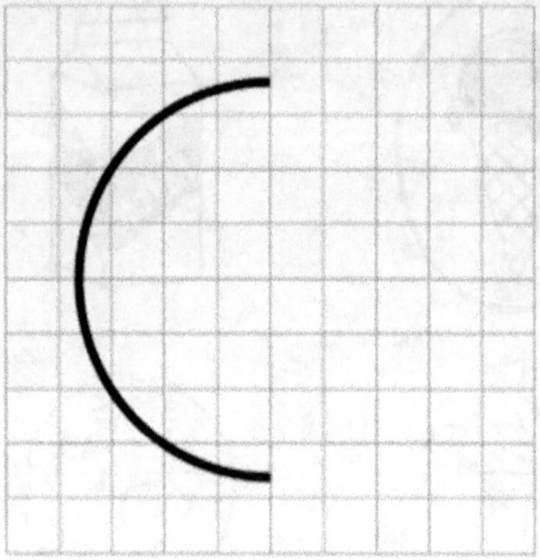

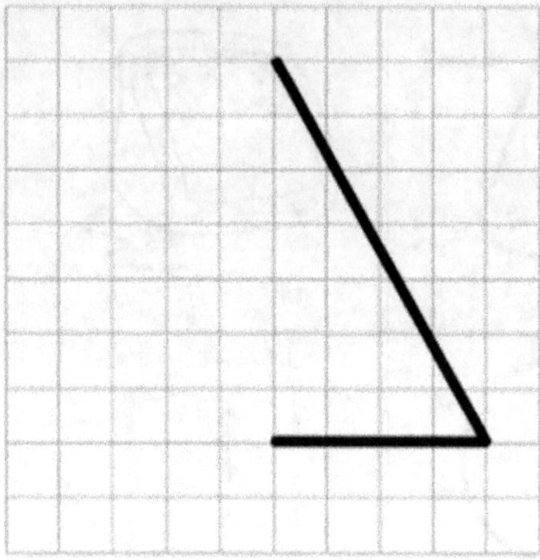

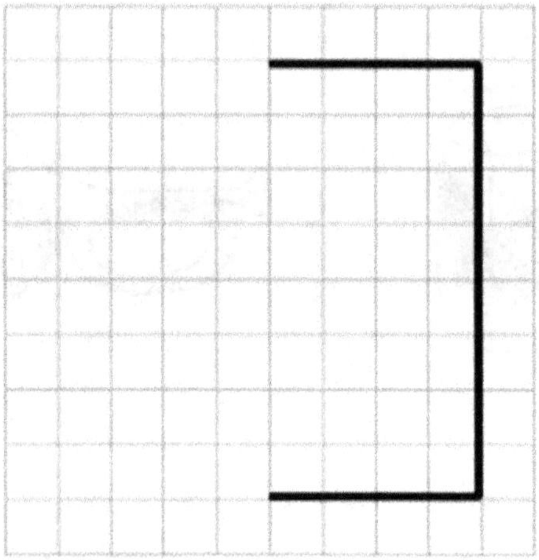

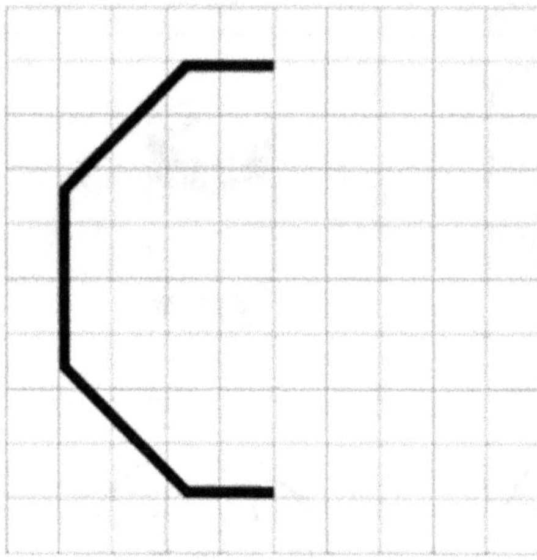

Workbook Answers

Chapter 1: Counting & Number Sense

Day 1

Place Value

Draw an X over the third and twelfth spoons.

Draw an X over the fourth, ninth, and thirteenth boxes.

Draw an X over the eighth, tenth, and fourteenth lamps.

Draw an X over the first, eleventh, and fifteenth bags.

Draw an X over the second, fifth, and ninth objects.

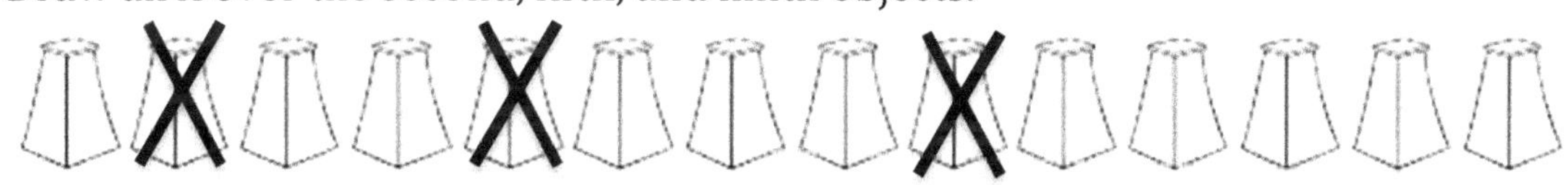

Draw an X over the fourth, seventh, and thirteenth globes.

Day 2

Greater Than, Less Than, or Equal: 2 Digit Numbers

1. <
2. >
3. <
4. >
5. >
6. <
7. <
8. =
9. <
10. <
11. >
12. <
13. <
14. >
15. <
16. <
17. =
18. =
19. >
20. <
21. =
22. >
23. <
24. >
25. <
26. <
27. =
28. >

Day 3

Rounding to the Tens

1. 40
2. 20
3. 90
4. 20
5. 30
6. 60
7. 80
8. 10
9. 70
10. 70
11. 30
12. 80
13. 100
14. 80
15. 10
16. 100
17. 80
18. 10
19. 60
20. 20
21. 60
22. 30
23. 80
24. 90
25. 20
26. 80
27. 70
28. 20

Rounding to the Nearest Hundreds

1. 400	15. 800
2. 500	16. 300
3. 100	17. 300
4. 700	18. 200
5. 200	19. 900
6. 400	20. 900
7. 500	21. 300
8. 200	22. 100
9. 200	23. 900
10. 100	24. 300
11. 600	25. 500
12. 300	26. 400
13. 200	27. 700
14. 800	28. 300

Day 4

Arrange Numbers from Smallest to Largest

1.	188	2	2.	879	3	3.	200	3
	13	1		233	2		756	4
	689	4		54	1		89	2
	346	3		95	4		2	1
4.	346	2	5.	421	1	6.	358	4
	745	3		465	2		101	3
	989	4		777	4		7	1
	124	1		758	3		45	2

Arrange Numbers from Largest to Smallest

1.	123	4	2.	564	2	3.	54	3
	676	2		2	4		123	2
	879	1		99	3		124	1
	333	3		612	1		53	4
4.	202	3	5.	99	2	6.	125	4
	457	2		45	4		141	3
	589	1		54	3		223	2
	2	4		470	1		789	1

Day 5

Understanding Place Value 1

1. 49
2. 20
3. 65
4. 55
5. 91

Understanding Place Value 2

1. 695
2. 173
3. 461
4. 229
5. 934
6. 318
7. 888
8. 571
9. 999
10. 415

Understanding Place Value 3

1. 749
2. 420
3. 515
4. 155
5. 391

Day 6

Writing Numbers in Word Form

1. Seventy-Six
2. Thirty-Four
3. Fifty-Nine
4. Thirteen
5. Ninety-Nine
6. Eighty-Five
7. Twenty-Four
8. Forty-Two
9. Four Hundred Thirty-Five
10. Two Hundred Eleven
11. One Hundred Forty-Seven
12. Six Hundred Fifty-Four
13. One Hundred Seventy-Four
14. Six Hundred Ninety-Seven
15. Three Hundred Eighty-Seven
16. Three Hundred Eleven

Day 7

Recognizing Patterns 1

1. 2, B

2. 2, B

3. C, F

4. 3, D

1. 7	9. 32
2. 3	10. 95
3. 9	11. 50
4. 21	12. 5
5. 77	13. 10
6. 15	14. 13
7. 18	
8. 46	

1. 13	9. 5
2. 9	10. 16
3. 18	11. 37
4. 53	12. 13
5. 4	13. 9
6. 78	14. 20
7. 42	
8. 64	

Day 8

Recognizing Patterns 2: Fill in the Number Chart

10	20	30	40	50	60	70	80	90	100
110	120	130	140	150	160	170	180	190	200
210	220	230	240	250	260	270	280	290	300
310	320	330	340	350	360	370	380	390	400
410	420	430	440	450	460	470	480	490	500
510	520	530	540	550	560	570	580	590	600
610	620	630	640	650	660	670	680	690	700
710	720	730	740	750	760	770	780	790	800
810	820	830	840	850	860	870	880	890	900
910	920	930	940	950	960	970	980	990	1000

Chapter 2: Addition

Day 9

Single-Column Addition

1.	7 + 2 = ⑨	3 + 5 = 8
2.	5 + 4 = ⑨	2 + 1 = 3
3.	5 + 5 = ⑩	3 + 1 = 4
4.	1 + 2 = 3	8 + 3 = ⑪
5.	5 + 2 = 7	9 + 5 = ⑭
6.	6 + 1 = 7	7 + 4 = ⑪
7.	2 + 0 = 2	4 + 7 = ⑪
8.	3 + 2 = 5	8 + 5 = ⑬
9.	1 + 8 = 9	9 + 5 = ⑭
10.	2 + 7 = 9	6 + 6 = ⑫
11.	8 + 9 = ⑰	4 + 8 = 12

Day 10

Single-Column Addition: Greater Than, Less Than or Equal

1. 4 < 12
2. 9 = 9
3. 12 > 11
4. 7 < 9
5. 10 > 7
6. 13 > 11
7. 9 > 2
8. 9 = 9
9. 16 > 9
10. 3 < 6
11. 12 < 14

Day 11

2-Digit Addition: Regrouping

1. 21
2. 40
3. 81
4. 62
5. 55
6. 31
7. 41
8. 74
9. 35
10. 33
11. 102
12. 83

Day 12

2-Digit Addition: Number Find

41	64	92	45
+ 8	+ 4	+ 7	+ 5
49	68	99	50
24	37	16	26
+ 3	+ 2	+ 1	+ 9
27	39	17	35
57	19	21	85
+ 7	+ 4	+ 7	+ 9
64	23	28	94

Day 13

Commutative Property

1. 90, 5
2. 71, 6
3. 10, 70
4. 8, 7
5. 21, 25
6. 24, 74
7. 7, 13
8. 51, 13
9. 5, 87
10. 9, 20
11. 62, 3
12. 31, 3

Day 14

2-Digit Addition

1. 35
2. 96
3. 39
4. 84
5. 99
6. 85
7. 84
8. 25
9. 88
10. 89
11. 99
12. 99

Day 15

3-Digit Single-Column Addition

1. 16
2. 7
3. 19
4. 11
5. 15
6. 12
7. 16
8. 17
9. 4
10. 23
11. 16
12. 11
13. 18

Day 16

4-Digit Single-Column Addition

1. 13

2. 12

3. 20

4. 12

5. 19

6. 18

Chapter 3: Subtraction

Day 17

Single-Column Subtraction: Greater Than, Less Than or Equal

1. 4 < 5
2. 5 > 2
3. 0 < 3
4. 2 > 1
5. 1 = 1
6. 8 > 2
7. 2 < 7
8. 1 < 7
9. 0 < 2
10. 5 > 3
11. 3 = 3

Day 18

2-Digit Subtraction

1. 73
2. 32
3. 51
4. 11
5. 20
6. 12
7. 24
8. 46
9. 80
10. 64
11. 83
12. 42

DAY 19

2-DIGIT SUBTRACTION: NUMBER FIND

41	11	16	89
- 1	- 2	- 9	- 3
40	9	7	86

21	3	23	77
- 3	- 1	- 10	- 5
18	2	13	72

12	61	39	18
- 6	- 1	- 9	- 6
6	60	30	12

DAY 20

2-DIGIT SUBTRACTION

1. 12
2. 12
3. 24
4. 11
5. 25
6. 2
7. 24
8. 86
9. 10
10. 18
11. 3
12. 22

DAY 21

2-DIGIT SUBTRACTION: BORROWING

1. 37
2. 35
3. 38
4. 18
5. 9
6. 21
7. 5
8. 9
9. 12
10. 6
11. 19
12. 53

DAY 22

WRITING OUT 2-DIGIT SUBTRACTION PROBLEMS

1. 16
2. 28
3. 18
4. 2
5. 13
6. 8
7. 3
8. 43

Day 23

Subtraction Word Problems

1. 2

2. 6

3. 4

4. 3

5. 63

Chapter 4: Fractions & Graphing

Day 24

Halving

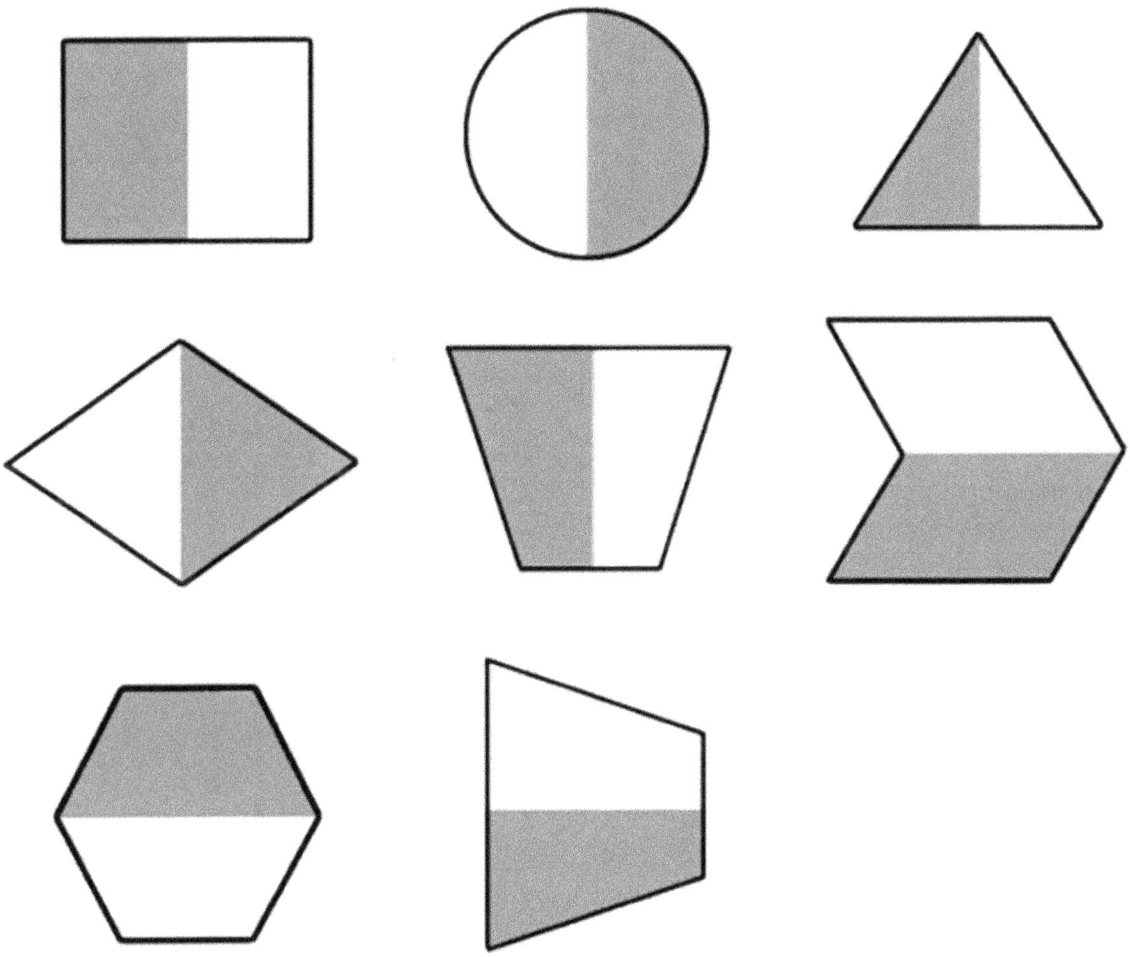

Day 25
Identifying Fractions

$\frac{1}{3}$ = 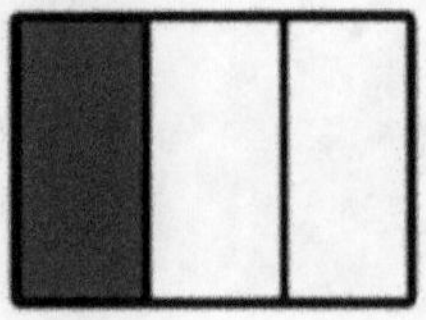

$\frac{2}{6}$ = 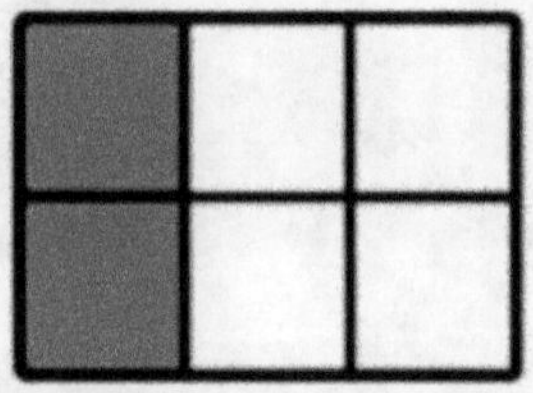

$\frac{1}{4}$ = 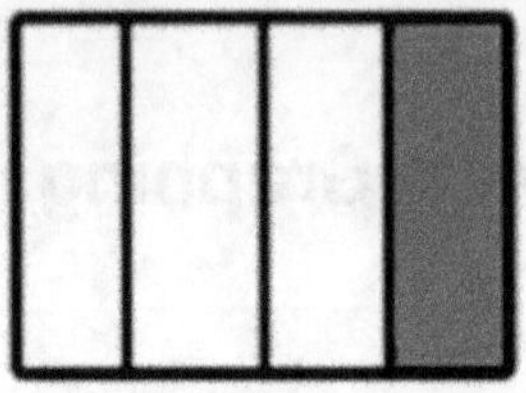

$\frac{5}{6}$ = 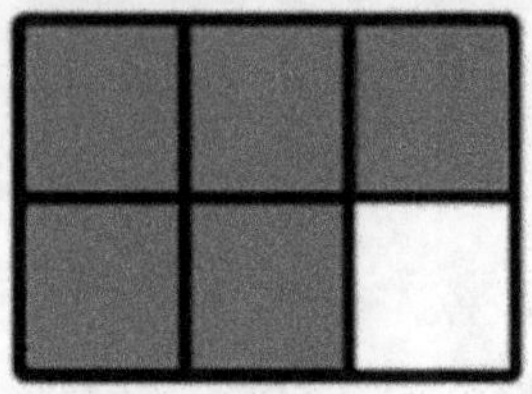

$\frac{2}{3}$ = 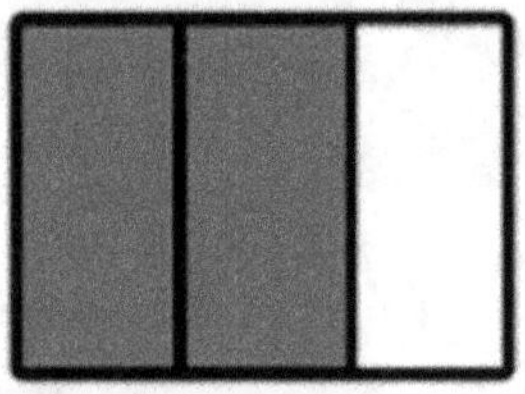

$\frac{7}{9}$ = 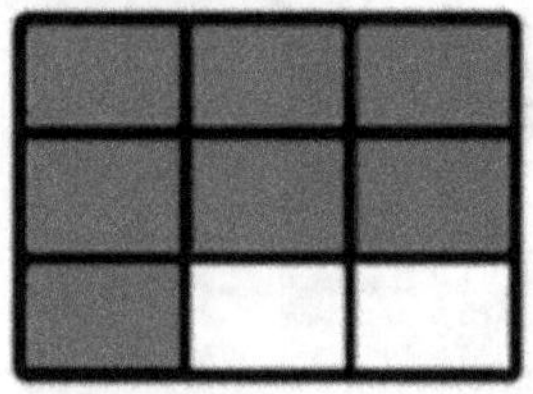

$\frac{3}{8}$ = 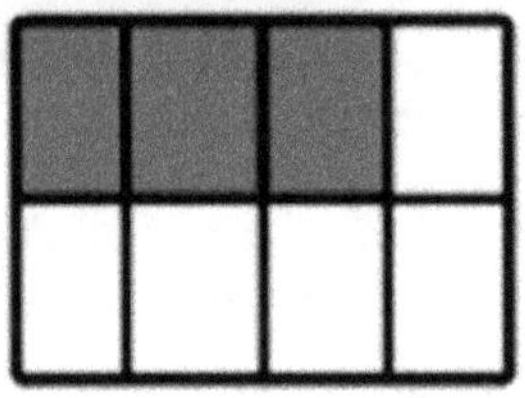

$\frac{1}{2}$ = 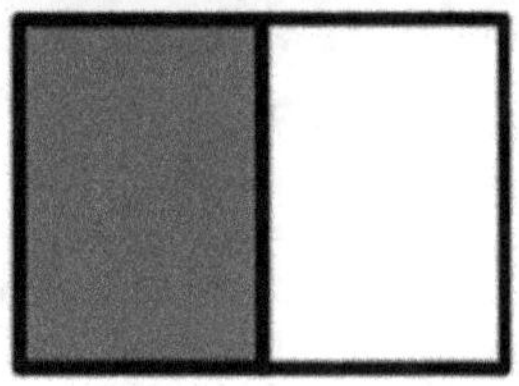

DAY 26

WRITING FRACTIONS

	=	$\frac{1}{2}$		=	$\frac{1}{3}$
	=	$\frac{1}{3}$		=	$\frac{1}{4}$
	=	$\frac{1}{4}$		=	$\frac{1}{2}$
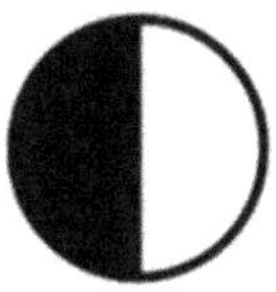	=	$\frac{1}{2}$		=	$\frac{1}{3}$

DAY 27
FRACTION WORD PROBLEMS

1. Jimmy ate half of his sandwich. Color half of his sandwich.

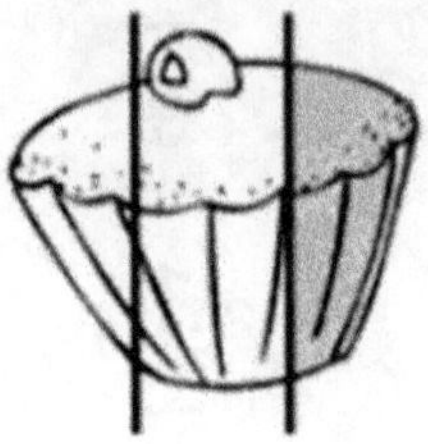

2. Amy can have one third of this cupcake. Color one third of her cupcake.

3. Penny ate one quarter of the pizza. Color one quarter of her pizza.

Day 28

Graphing

Workbook Answers

Day 29
Graphing: Identifying Coordinates

Across	Up	Across	Up	Across	Up	Across	Up
3 ,	4	5 ,	6	6 ,	1	2 ,	2

Day 30
Graphing Word Problems

1. 3

2. Sally

3. Tim

4. 5

5. Lily: 8, Dave: 6, Sally: 10, Tim: 5

Chapter 5: Counting Money

DAY 31

COUNTING CHANGE

	1	1	0	3
	2	1	1	
	2	3	1	2

Workbook Answers

Day 32
Cents and Dollars

How many  = 

4

How many  = 

100

How many  = 

20

How many  = 

10

Day 33

Dollars and Larger Bills

How many  =  **5**

How many  =  **10**

How many  =  **2**

How many  =  **4**

Day 34
Dollars: Greater Than, Less Than or Equal

<

=

>

>

Day 35

Dollars: Matching

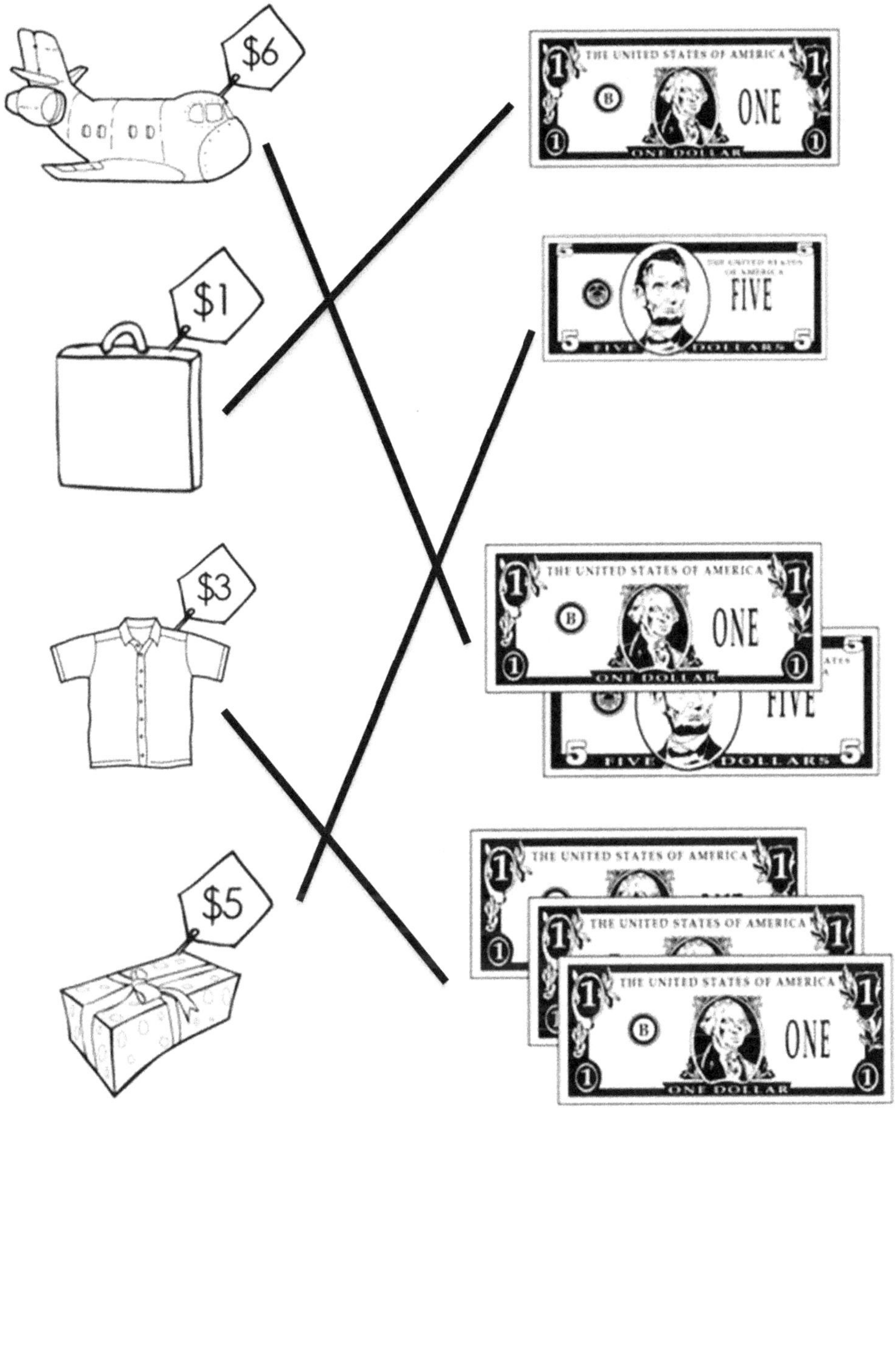

Chapter 6: Time

DAY 36

IDENTIFYING TIME

3:00

9:00

5:00

7:00

3:30

9:30

5:45

6:15

6:00

10:00

7:00

11:00

12:15

1:30

4:15

2:45

DAY 37

WRITING TIME

Day 38

Drawing Time

DAY 39

TIME MATCHING

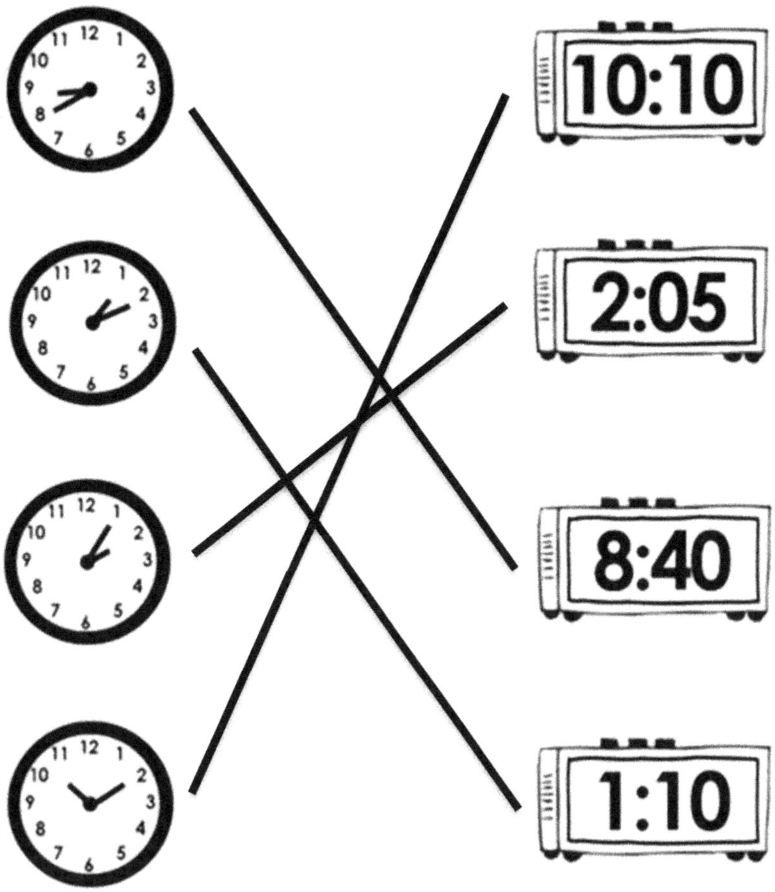

Day 40

Writing Time Passage

What time will it be in 3 hours and 30 minutes?

What time will it be in 6 hours and 30 minutes?

What time will it be in 1 hour and 45 minutes?

What time will it be in 3 hours and 30 minutes?

What time will it be in 3 hours and 45 minutes?

What time will it be in 2 hours and 30 minutes?

Chapter 7: Measurements

DAY 41

IDENTIFYING LIQUID MEASUREMENTS

2

1

3

4

3

2

4

1

3

2

1

4

Workbook Answers

DAY 42
COMPARING LIQUID MEASUREMENTS

 or 

 or 

 or 

 or 

Day 43

Identifying Warmer Temperatures

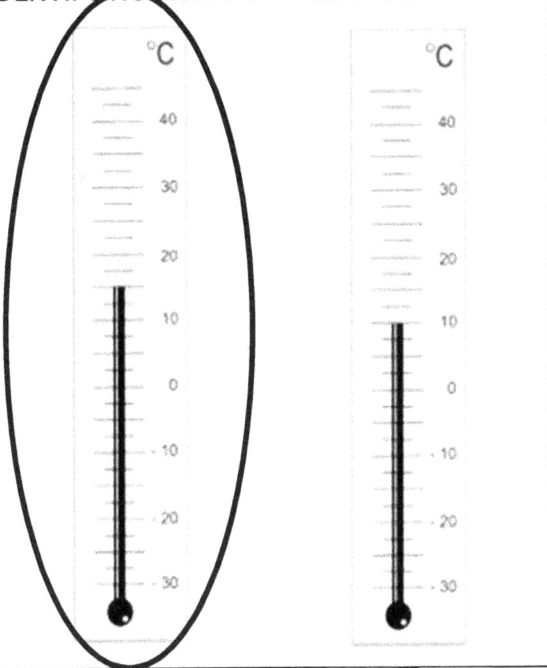

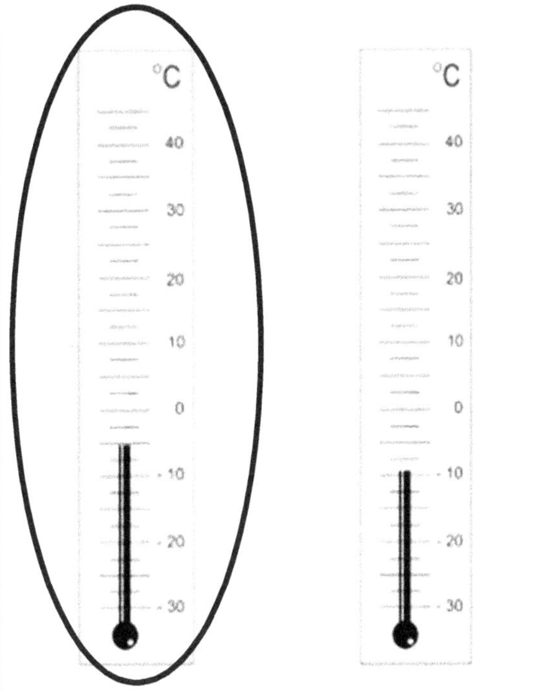

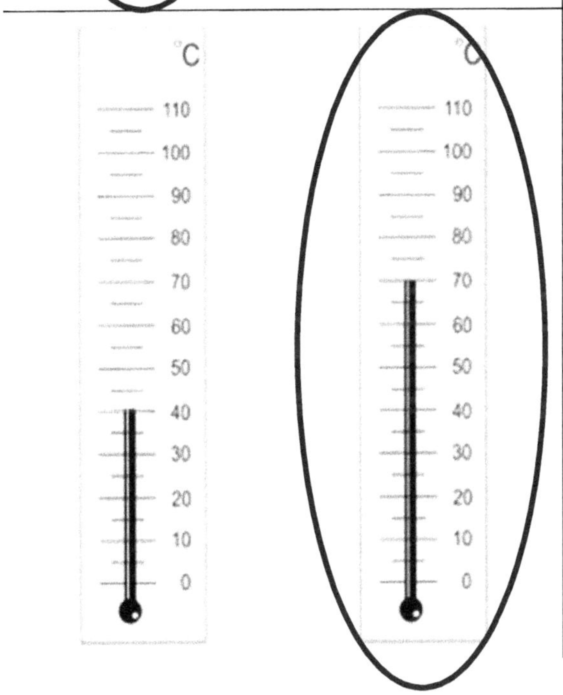

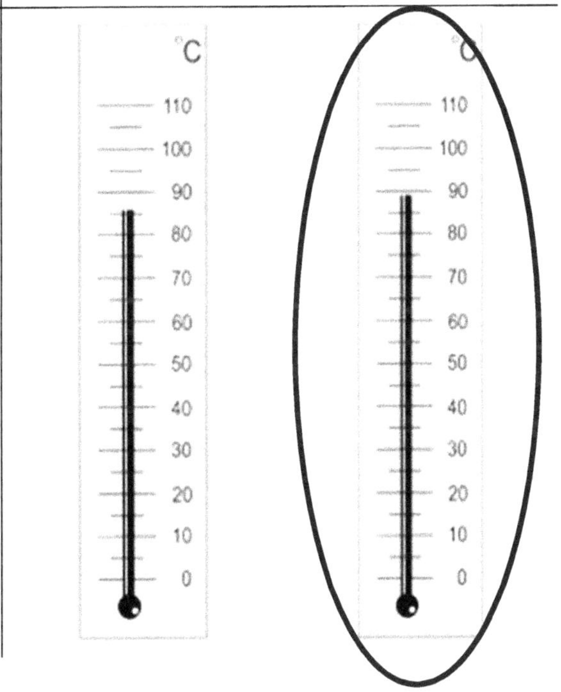

Day 44
Writing Temperature

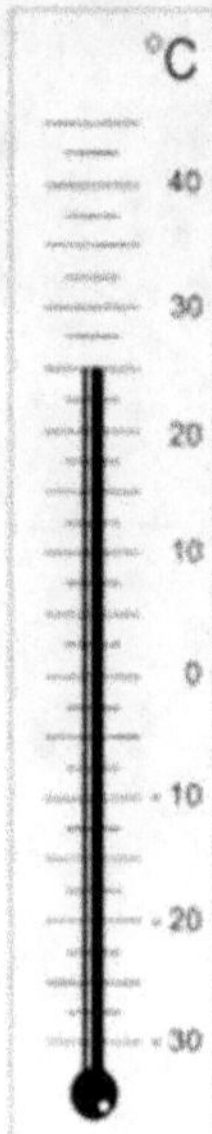

25°

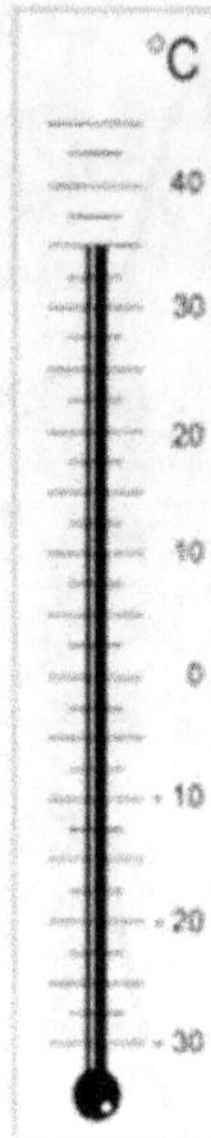

35°

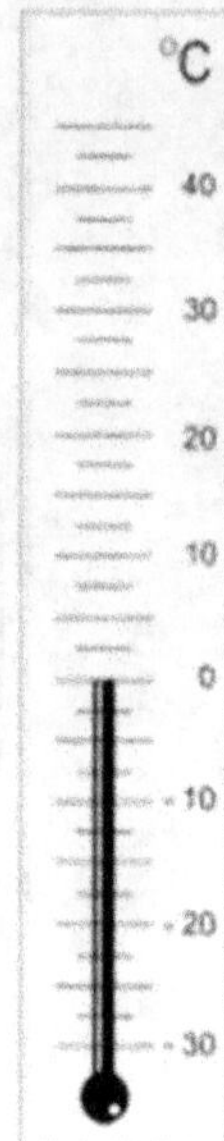

0°

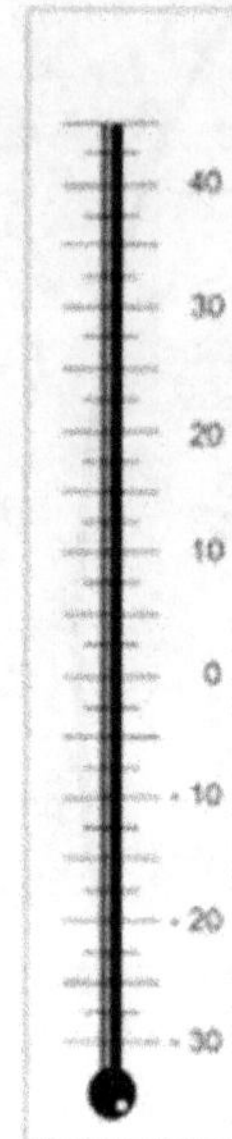

45°

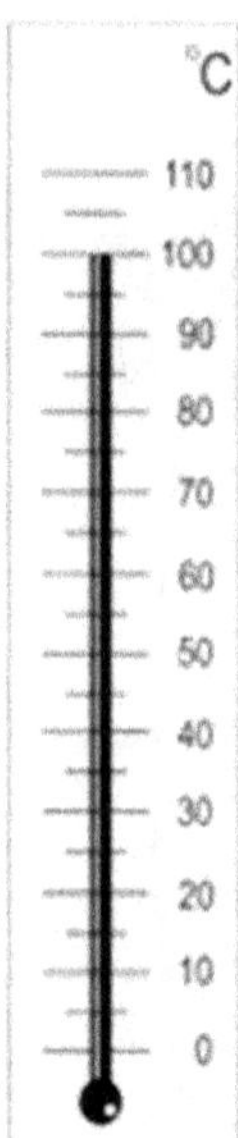

100°

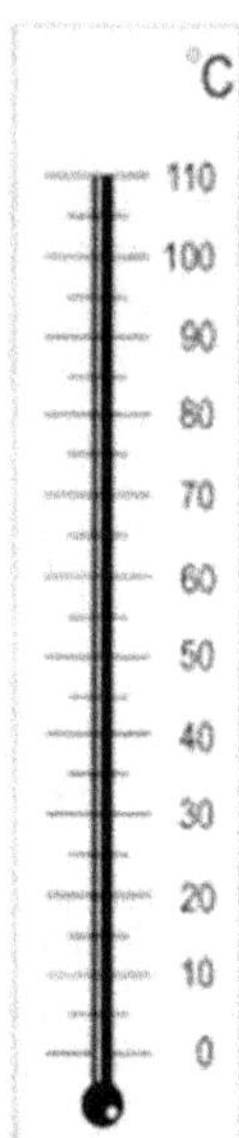

110°

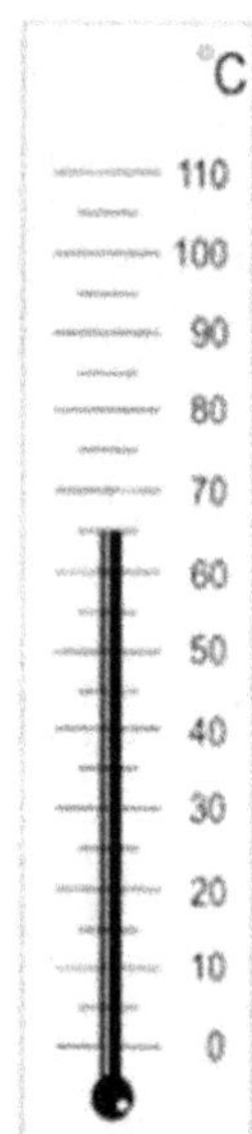

65°

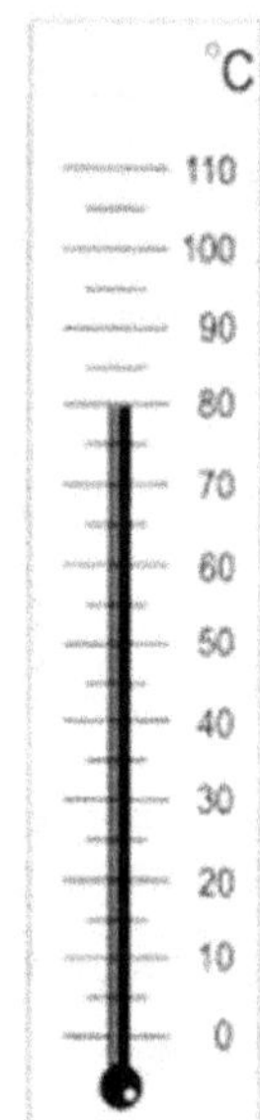

80°

DAY 45

IDENTIFYING LENGTH IN CENTIMETERS

15 centimeters	17 centimeters	12 centimeters
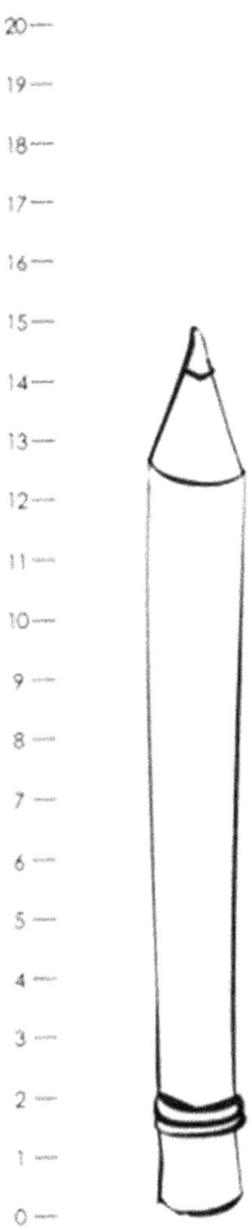	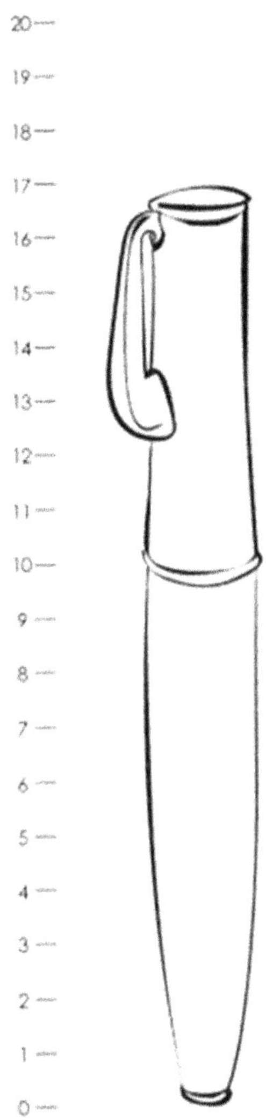	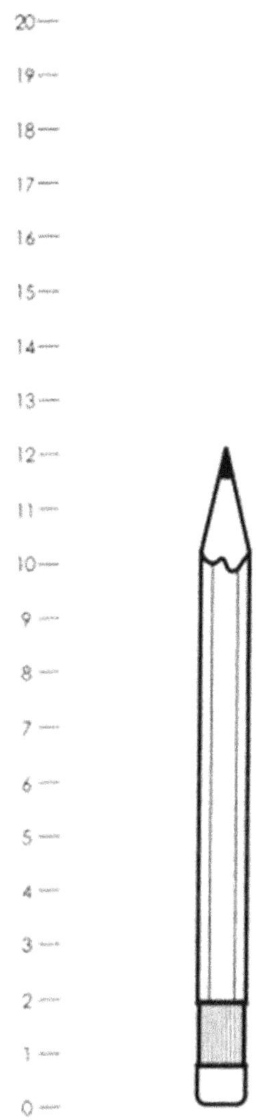

Chapter 8: Geometry

Day 46

Identifying Flat Shapes

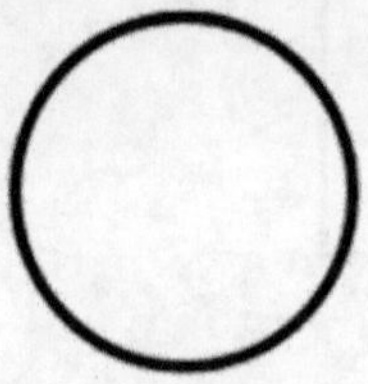

Circle

Rectangle

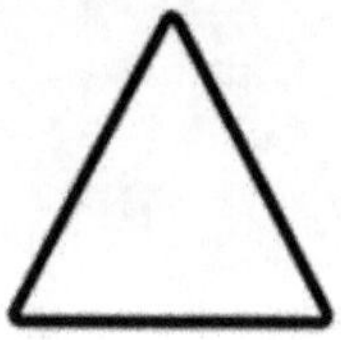

Triangle

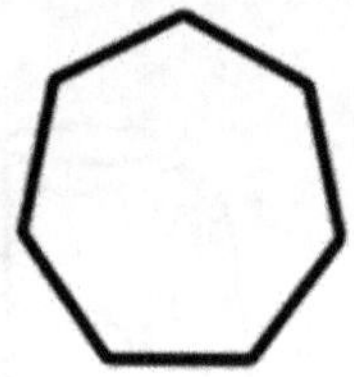

Heptagon

Pentagon

Square

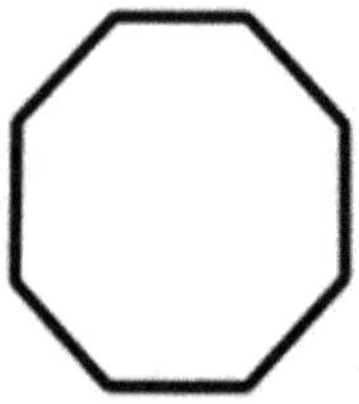

Octagon

Hexagon

Day 47

Identifying Solid Shapes

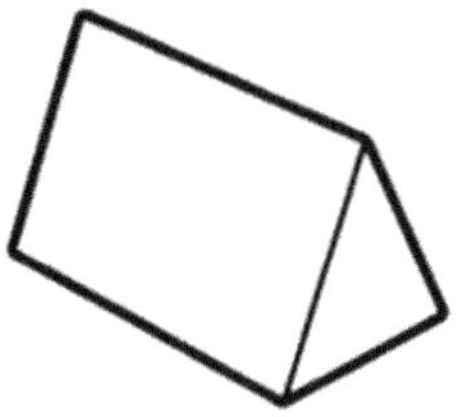

Prism

Cone

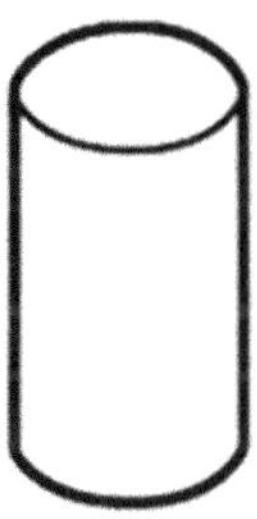

Cylinder

Sphere

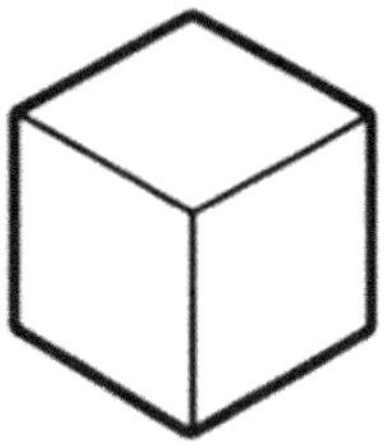

Cube

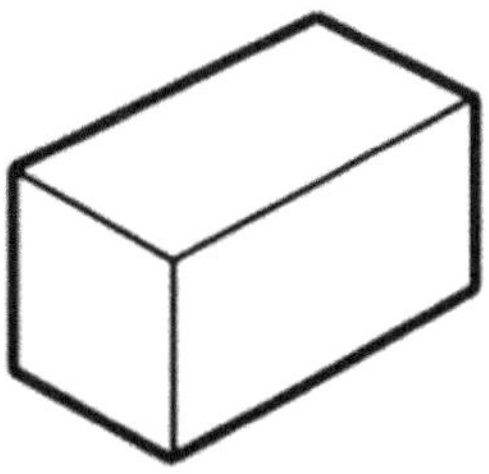

Rectangular Prism

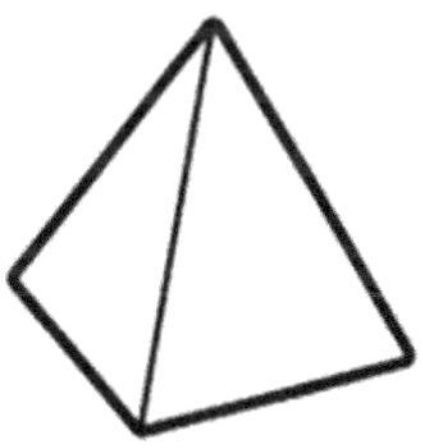

Pyramid

DAY 48
MAKING NEW SHAPES

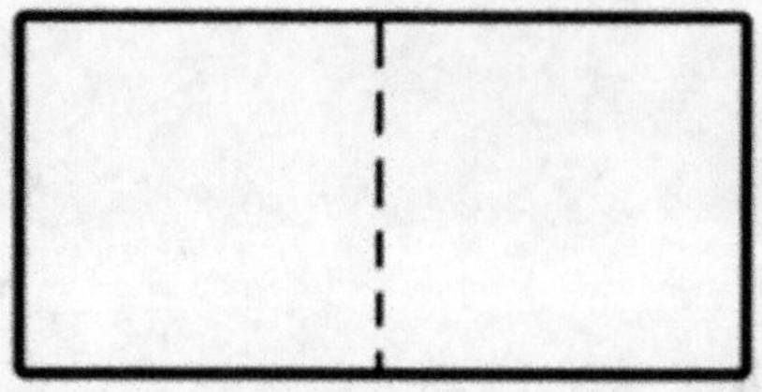

1. Make two squares.

2. Make a square and a triangle.

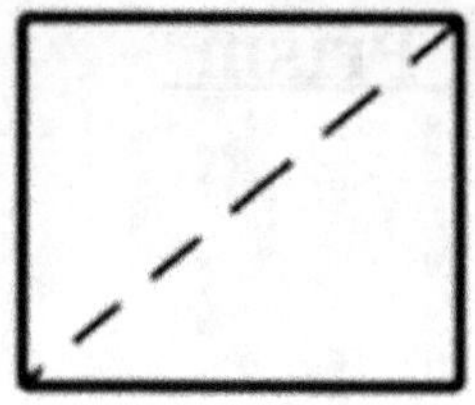

3. Make two triangles.

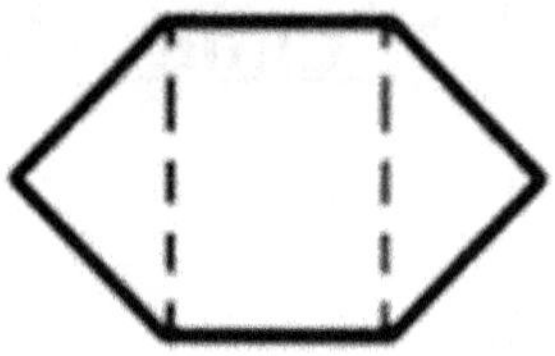

4. Draw two lines to make a square and two triangles.

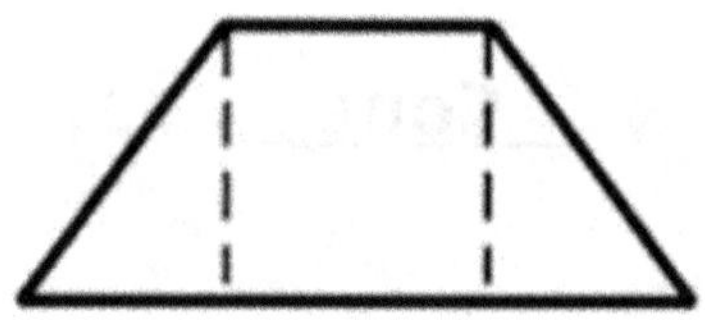

5. Draw two lines to make a square and two triangles.

Day 49

Comparing Solid Shapes

Cone

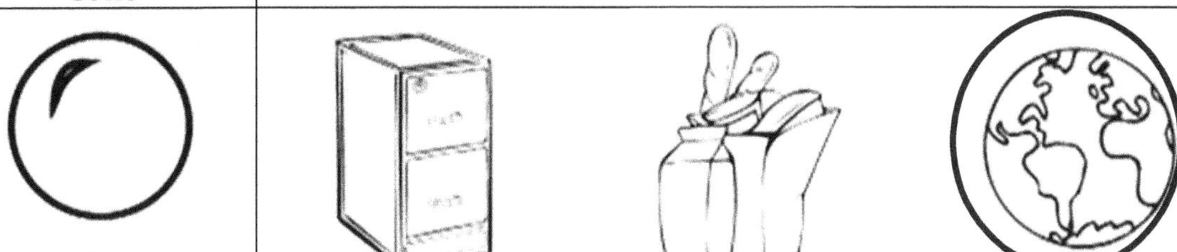

Sphere

Cylinder

Pyramid

DAY 50
SYMMETRY

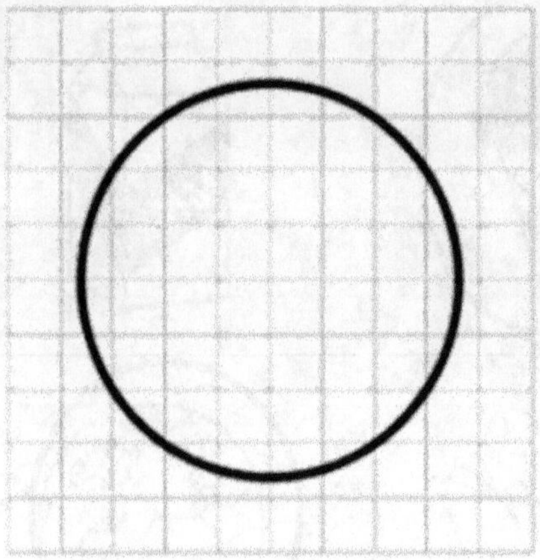
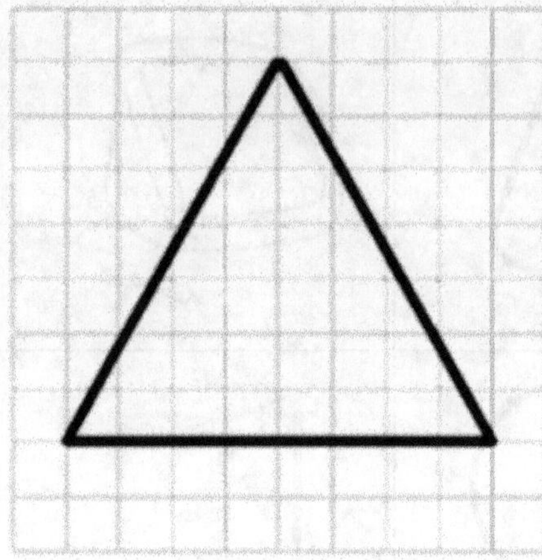
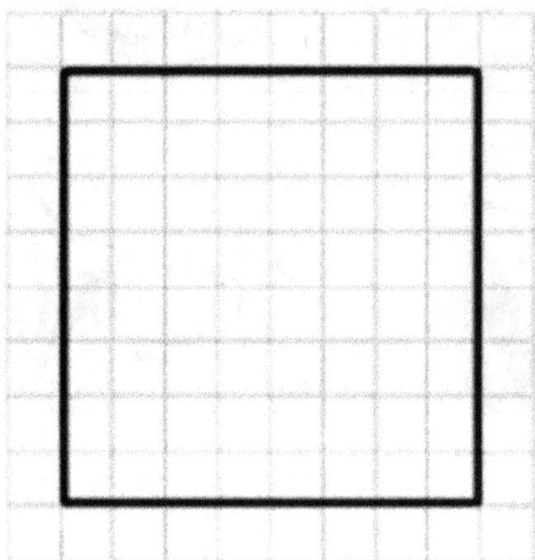
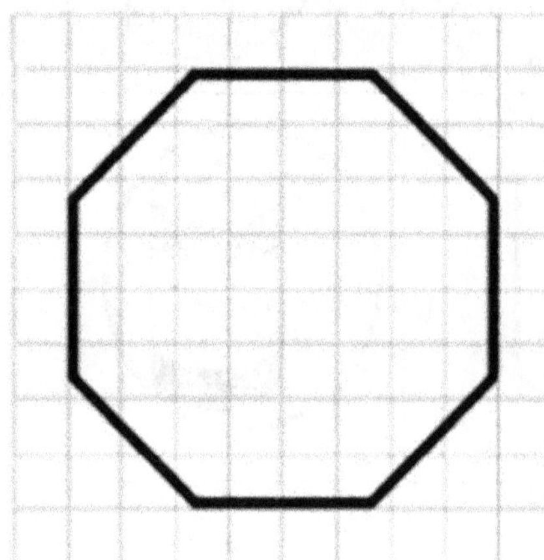

www.ingramcontent.com/pod-product-compliance
Lightning Source LLC
LaVergne TN
LVHW082355200926
848481LV00012B/3568